THE SCIENCE OF LIFE AND DEATH IN
FRANKENSTEIN

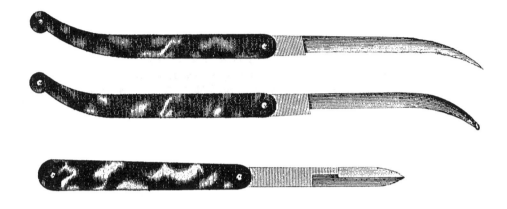

THE SCIENCE OF LIFE AND DEATH IN
FRANKENSTEIN

SHARON RUSTON

BODLEIAN
LIBRARY
PUBLISHING

First published in 2021 by the Bodleian Library
Broad Street, Oxford OX1 3BG
www.bodleianshop.co.uk

ISBN 978 1 85124 557 4

Text © Sharon Ruston, 2021
All images, unless specified, © Bodleian Library, University of Oxford, 2021

Sharon Ruston has asserted her right to be identified as the author of this Work.

All rights reserved.

No part of this book may be reproduced, stored in a retrieval system, or transmitted in any form or by any means, electronic, mechanical, photocopying, recording, or otherwise, without the written permission of the Bodleian Library, except for the purpose of research or private study, or criticism or review.

Publisher: Samuel Fanous
Managing Editor: Deborah Susman
Editor: Janet Phillips
Picture Editor: Leanda Shrimpton
Production Editor: Susie Foster
Cover design by Dot Little at the Bodleian Library
Designed and typeset by Lucy Morton of illuminati in 12 on 16 Fournier
Printed and bound by Livonia Print, Latvia, on 90 gsm Munken Premium Cream paper

British Library Catalogue in Publishing Data
A CIP record of this publication is available from the British Library

CONTENTS

	ACKNOWLEDGEMENTS	vii
	INTRODUCTION	1
ONE	**LIFE AND DEATH IN ROMANTIC LITERATURE**	19
TWO	**VITAL AIR**	40
THREE	**ELECTRIC LIFE**	61
FOUR	**THE VITAL PRINCIPLE**	81
FIVE	**RAISING THE DEAD**	107
	AFTERWORD	121
	NOTES	125
	FURTHER READING	140
	PICTURE CREDITS	147
	INDEX	149

ACKNOWLEDGEMENTS

I had so much fun writing this book. On a sabbatical from my teaching at Lancaster University, I was awarded a Fellowship at St Catherine's College, Oxford, and wrote most of the book in the Bodleian Library. Thanks to Kristen Sheppard-Barr and Jyotsna Singh for the time I spent in Oxford. I am very grateful to Stephen Hebron for showing me the *Frankenstein* manuscripts and Mary Shelley's journals, which are held at the Bodleian Library, and for the support and advice of Samuel Fanous, Janet Phillips and Leanda Shrimpton at Bodleian Library Publishing. But this book is also the result of many years of teaching *Frankenstein*. Thanks especially to students, past and present, on my third-year special subject module 'Monstrous Bodies' at Lancaster University. I would like to give particular thanks to Julie Sheppard and Sara Cole for reading the whole book and giving me such useful feedback.

Many people have read a chapter of the book and everyone who has done so has made excellent suggestions and contributed to the whole. Others responded to individual queries. I am grateful to Simon Bainbridge, Lou Berriman, Jo Carruthers, Sue Chaplin, Jon Dickson, Jan Golinski, Nick Groom, Daisy Hay, John Holmes, Sam Illingworth, Jelena Lee, Francesca Locke, Andrew Moor, Iwan

Rhys Morus, Frank Pearson, Catherine Spooner, James Secord, Dale Townshend, Dale Watson, David Wootton, Angela Wright. Any errors are of course my own. Finally, the book is dedicated to the love of my life, Jerome de Groot.

INTRODUCTION

Mary Shelley experienced more than her fair share of death.[1] Her mother, Mary Wollstonecraft, died only eleven days after giving birth to her, and she had to live with the fact that her birth had inadvertently caused her mother's death. She grew up without a mother to help and guide her through life, just as the Creature in *Frankenstein* is abandoned by his creator. Thereafter, deaths quickly followed births in a horribly predictable fashion in Mary's life. She eloped, aged 17, with Percy Bysshe Shelley: he was 22, already married and a published poet. Their brief life together was fraught with bereavements, money worries, fears that he would be imprisoned for debt and anxieties about his health. Life and death were subjects on Mary's mind throughout this time.

Back in England after eloping to the Continent, Mary gave birth to her first child, who was born prematurely on 17 February 1815 and died on 6 March. She was successfully delivered of William on 24 January 1816, and he was with them for the momentous trip to Geneva when Mary conceived *Frankenstein*. Mary's stepsister Claire Clairmont had accompanied the Shelleys when they eloped, and in Geneva the Shelleys and Claire spent their days and nights with Lord Byron and his doctor, John Polidori. On their return to England,

Claire gave birth to Allegra, Lord Byron's child, on 12 January 1817. But news of death soon followed. On 9 October 1817, Mary's half-sister Frances Imlay Godwin (known as Fanny), the daughter of Mary Wollstonecraft and Gilbert Imlay, committed suicide by overdosing on laudanum. A few months later, Mary and Percy heard that Harriet Shelley, Percy's first wife, had killed herself. She had been found drowned in the Serpentine, pregnant, on 10 December. Mary and Percy married soon after, on 30 December, but Percy was refused custody of his two children by Harriet on the grounds that his publications were immoral.

Mary was pregnant during the composition of *Frankenstein*, which she finished just before she gave birth to Clara Everina, on 2 September 1817. She referred to the novel as her 'hideous progeny' and 'offspring' and it focuses on issues of creation, life and death.[2] She could not have known at this point that there was so much sadness and loss still left to experience. Clara died on 24 September 1818, at only a year old, and William on 7 June 1819, aged 3. The Shelleys' only child to survive, Percy Florence, was born on 12 November 1819. Claire and Byron's daughter Allegra died on 22 April 1822, aged 5. In June 1822 Mary had a near-fatal miscarriage. Percy drowned on 8 July 1822. By any measure, Mary experienced more bereavement than most.

Mary Shelley's novel *Frankenstein* emerged from personal experience and cutting-edge science. Life, for Mary and her contemporaries, seemed a tenuous state of existence and death an ever-present possibility. As scientific knowledge increased and new resuscitation techniques were widely reported, the public worried that the distinction between life and death was not as clear as had been thought. There was a real concern that doctors could not tell with any precision when very ill people were alive or dead. People worried that they might be buried alive or their corpses robbed from their graves for use in medical experiments. This book reveals how Mary Shelley exploited the uncertainty caused by the new scientific and medical ideas of life and death, and explains why *Frankenstein* continues to produce disquiet and unease 200 years after its first publication.

THE LIFE AND DEATH OF MARY SHELLEY'S FIRST CHILD

Even before she had thought of her novel, Mary had experienced birth and death, and the slippery divide between them. She was very unwell during her first pregnancy, complaining of 'a very bad side ache' on Christmas Day in 1814. She consulted a doctor, called 'Lawrence', on 26 December.[3] On 5 January 1815, she was 'very ill'; on 12 January she was 'very unwell' and on 6 February she was 'not well' and went to bed early. Perhaps Mary had an inkling that something was not right with her pregnancy given the state of her health. For two months she recorded feeling unwell in her journal. At the same time, she noted that Percy was 'very ill' at the end of January, and on 14 February she wrote 'All unwell in the evening'.[4] The entire household was ill. A few days later, on 17 February 1815, Mary gave birth.

Percy recorded Mary's labour in the journal they shared, writing that it was accompanied by 'very few addit[i]onal pains'. He noted that 'Dr Clarke' had arrived five minutes after the birth. Extraordinarily, he may have been the brother of the doctor who attended her mother's disastrous labour. Percy wrote in their journal: 'all is well. Maie [a pet name for Mary] perfectly well & at ease The child is not quite 7 months. The child not expected to live. S. [Percy] sits up with Maie. much agitated & exhausted.'[5] It is baffling that such a situation could be summed up in his beginning with 'all is well'.

The slide from birth to expected death was fast, as was the change in Mary being considered 'at ease' to being 'agitated & exhausted'. The days thereafter were marked by the child's persistent clinging to life. The next day, Percy wrote, 'The child unexpectedly alive, but still not expected to live.' On Friday 24 February the child exhibited 'favourable symptoms' and the parents began to 'indulge' in 'hopes', which Dr Clarke confirmed. While the child and Mary continued to be well, though, Percy was 'very unwell' and then 'unwell & exhausted', culminating on 26 February, the first day that Mary rose from her bed, with 'S. has a spasm'.[6] Percy would have these convulsive fits throughout their life together, often in times of stress. He traced their source to an episode at Tan-yr-Allt, Wales, in 1813, when he claimed to have been injured by an intruder.

Despite seeming improvement and entries in the journal that record Mary nursing the baby, the devastating journal entry for 6 March reads: 'find my baby dead'. The effect on Mary's mental health is clear: she was 'not in good spirits'.[7] She wrote immediately to Thomas Jefferson Hogg, Percy's friend, and told him that she needed him, rather than Percy, because 'you are so calm a creature & Shelley is afraid of a fever from the milk – for I am no longer a mother now'.[8] That evening, there was 'a fuss', which resulted in a late night: 'to bed at 3'.[9] Percy was worried about so-called 'milk fever', which supposedly began a few days after mothers had begun to lactate, as we know Mary had recently. Mary's breast milk had built up but there was no baby to feed. In this instance, becoming a mother and losing a child had the potential to endanger her own life.

The next days of her journals contain many heart-wrenching entries: 'still think about my baby – 'tis hard indeed for a mother to loose a child'.[10] In her letter to Hogg she gave more information on the circumstances of the baby's death. Apparently, she had risen in the night to nurse her but did not because she appeared to be sleeping so soundly. She realized with hindsight that her baby had been already dead at that point and added that 'from its appearance it evidently died of convulsions'.[11] Throughout the entries, Mary most often referred to her new daughter as 'my baby', while Percy speaks of 'the child'. But Mary also does not gender the baby female in the journal, instead always referring to 'it', and it seems that the child was never given a name. This offers a strange prefiguring of the Creature in *Frankenstein*, who, while being gendered male in the novel, is often thought of as an 'it' by others, and is never named.

Looking back over the journal entries it is clear that the baby's frail and vulnerable start in life had seemed to improve before life was cruelly and suddenly taken away. The boundary between life and death is indeed thin in such cases and the ease with which a new life can be lost is troubling. Immediately after the death of her child Mary began to read Ovid's *Metamorphoses* in Latin; she continued throughout April and May, noting the number of pages she 'construes' each day. In *Metamorphoses* there are many instances of figures being transformed, just before death, into other forms with new lives. Changes in states of existence seem to have been on her mind, since she ended this journal around this time with: 'I begin a new journal with our regeneration'.[12]

This period, so well documented in Mary's journals, gives us an insight into how much the Shelleys' lives were consumed by poor health. Years later, Mary recorded in a note that around this time Percy had been diagnosed as 'dying rapidly of a consumption' but that treatment advised by their doctor, William Lawrence, made him feel much better.[13] In 1817 he was again consulting Lawrence, who advised him to stop writing poetry because it excited his nerves too much, and to travel to the warmer climate of Italy. Throughout his life, Percy suffered from periods of intense pain, which were diagnosed at different times as hepatitis, nephritis and opthalmia. Perhaps his own physical suffering partly explains his lifelong interest in medicine, having briefly considered it as a profession after he was expelled from the University of Oxford. Lawrence's edict was the reason that the Shelleys travelled abroad again after their elopement. In many ways he was the reason behind that fateful trip to Geneva.

THE SCIENCE OF THE TIMES

The most immediate inspiration for Mary's novel occurred in mid-June 1816, in Villa Diodati, Geneva, the house that Byron and Polidori had rented, while the Shelleys and Claire Clairmont rented another nearby. The weather was awful and as a consequence they were forced to stay indoors and amuse themselves by telling ghost stories.[14] One night, as Mary Shelley remembered it years later, she sat listening to Byron and Percy discuss 'the nature of the principle of life, and whether there was any probability of it being discovered and communicated'.[15] At the time, Polidori wrote in his diary for 15 June, '[Percy] Shelley and I had a conversation about principles, – whether man was to be thought merely an instrument.'[16] The subject was a topical and controversial one and it sparked Mary's famous creation.

The exact topic of these conversations could have ranged over a number of fascinating new developments in science and medicine. In her 1831 introduction to the novel, Mary mentions galvanism (*see Chapter 3*), which was then understood as a kind of unique electricity that living animals possessed. Her journal reveals that she had attended a lecture given by André-Jacques Garnerin 'on Electricity – the gasses' on 28 December 1814.[17] Also in this introduction, Mary refers to spontaneous generation (*see Chapter 4*), the idea that microscopic animalcules seemed to emerge from nothing in rotting meat, though she makes light of the idea and seems to accept that life could not really materialize this way.

Debate on the nature of life was raging at precisely this time between John Abernethy and William Lawrence in the Royal College of Surgeons (*see Chapter 4*). Abernethy made claims for electricity being the vital principle, while others claimed it was oxygen or blood. The question of what life was consumed the scientific and medical worlds in Britain and spilled out into the public sphere. Polidori was particularly well qualified to speak on the topic. He had studied medicine at the University of Edinburgh and written his third-year dissertation on sleep disorders (*see Chapter 1*). Such studies led people to believe that animation could be 'suspended', that there were states of being between life and death. Certainly, Frankenstein boasts that life and death were to him merely 'ideal bounds', which he found a way to 'break through'.[18]

THE ROYAL HUMANE SOCIETY

We can glean evidence of Mary Shelley's medical knowledge from the novel itself. The Humane Society, which became the Royal Humane Society in 1787, was set up to promote advice to the public on how to revive people who had experienced a life-threatening accident (*see Chapter 2*). The Society thought the treatments it advocated were particularly suited to those who had drowned or suffocated. The success of their work in resuscitating people from near-death fostered new anxiety about being buried prematurely, and this frightening situation was exacerbated by the kinds of advice being disseminated by medical professionals. It was widely agreed that there were two kinds of death: 'incomplete' and 'absolute'. If you were unlucky enough to experience the latter, there was a real chance that your body would be dug up and sold by grave-robbers for medical demonstration after you had been buried (*see Chapter 5*).

There is a tangible link between Mary Wollstonecraft and the Royal Humane Society. While still grieving his wife's death in 1798, William Godwin published memoirs of her life that were frank and shocking to the point of scandal. He mentioned Wollstonecraft's two suicide attempts, in June and October 1795. Her lover, Gilbert Imlay, had proved unfaithful and refused to set up home with her. She was deeply unhappy. Wollstonecraft told Imlay that she was 'buried alive' in a 'living tomb' without him. In the letter that accompanied her second suicide attempt she wrote: 'If I am condemned to live longer, it is a living death.'[19] On this occasion, she threw herself off Putney Bridge into the River Thames. She was saved by strangers. A letter written to Imlay suggests that she was revived using Royal Humane Society methods: 'I have only to lament, that, when the bitterness of death was past, I was inhumanly brought back to life and misery.'[20] Before the event, she had been dreading this outcome: 'I go to find comfort, and my only fear is, that my poor body will be insulted by an endeavour to recall my hated existence.'[21] We can only wonder what means they tried in the effort to bring her back to life.

Mary Shelley's knowledge of Royal Humane Society techniques can be seen in *Frankenstein* and in her personal journal. The entry for 19 March 1815, after her first baby had died, reads: 'Dream that my little baby came to life again – that it had only been cold & that we rubbed it by the fire & it lived'[22] (*see Plate 5*). This is exactly what the Royal Humane Society and others recommended in such cases. They also claimed that such measures had 'very great hopes of success'.[23] The much-publicized successes of the Royal Humane Society made deaths all the more poignant. The society suggested that it was within an individual's power to save victims. Percy Shelley mentioned the Royal Humane Society in notes to his poem *Queen Mab*: 'the Humane Society restores drowned persons, and because it makes no mystery

FIG. 1 Illustration of rescue from drowning under ice in Hyde Park, Royal Humane Society, 1821.

of the method it employs, its members are not mistaken for the sons of God.'[24] His tone here is scientific: Percy favours the Society's methods over the miracles that Jesus allegedly performed when he raised people from the dead. The latter are shrouded in mystery whereas the Royal Humane Society openly promoted its methods for the recovery of those apparently drowned. Despite this, many people at this time thought there was something sacrilegious about seeming to raise the dead.

With the Royal Humane Society proclaiming its successes so loudly and widely, did people feel responsible when they could not save their loved ones? The Society claimed that they were able to bring people back from the dead. Percy seems to share this idea when he wrote in

INTRODUCTION   9

1819 of Dr Bell, who had attended the Shelleys' son William during his last illness: 'By the skill of the physician he was once reanimated after the process of death had actually commenced, and he lived four days after that time.'[25] Where we would say that poor William took a little longer to die than was originally predicted, Percy seems to think that he had died and been brought back to life temporarily.

Although the word 'resuscitation' was available to Mary Shelley to use at this time, in *Frankenstein* she tends to use a language of life being lost and restored. There are a number of occasions in the novel when characters faint; losing consciousness was widely considered a state of suspended animation. When Victor Frankenstein creates the Creature, he collapses and is described as 'lifeless'. In this instance his friend Henry Clerval 'restored [him] to life'. Elizabeth Lavenza faints on seeing the corpse of William Frankenstein: 'She fainted, and was restored with extreme difficulty. When she again lived, it was only to weep and sigh.'[26] This language suggests that while Elizabeth is unconscious, she is deemed to be dead. This is the language of the Royal Humane Society, which claimed that because of its methods 'upwards of a hundred and fifty persons, may with the strictest propriety, be said *to have been raised from the dead*'.[27]

THE SHELLEYS' KNOWLEDGE OF SCIENCE AND MEDICINE

Mary's education in science and medicine began before she married Percy Shelley, because her parents were Wollstonecraft and Godwin.[28] She grew up in a household that was interested in science and medicine; though she never knew her mother, she could recognize Wollstonecraft's grasp of these topics by reading her work. Wollstonecraft claims in her most famous work, *A Vindication of the Rights of Woman*, 'I have conversed, as man with man, with medical men, on anatomical subjects.'[29] She thought women 'should be taught the elements of anatomy and medicine' and even that 'Women might certainly study the art of healing, and be physicians as well as nurses.'[30] Through her published work, Wollstonecraft taught her daughter that these subjects were fit for women.

Decades later, when people learned that she had written *Frankenstein*, Mary recalled being asked repeatedly, 'How I, then a young girl, came to think of, and to dilate upon, so very hideous an idea?'[31] When *Frankenstein* was published, some people thought Percy had written it, not least because of its subject matter, which was not considered an appropriate conception for a girl of 18. The novel's central idea, that a living being could be artificially created and animated, was a live scientific issue at the time. How could a young girl have the necessary knowledge to write such a book? There were still no female physicians, as Wollstonecraft had hoped there would be in the future, but *Frankenstein* broached the science of life and death with confidence and insight. Its topicality is demonstrated by one contemporary reviewer declaring that the novel had 'an air of reality attached to it, by being connected with the favourite projects and passions of the times'.[32]

Wollstonecraft's scientific interests can be seen in the number of important books on natural history she reviewed, such as William Smellie's *Philosophy of Natural History* in the *Analytical Review* of October 1790.[33] Following her mother's lead, Mary read Smellie in October 1814.[34] The book describes humans, animals and plants: their similarities and their differences. In her review of Smellie's book, Wollstonecraft notes that he wished particularly to see '*How far peculiarities of structure are connected with peculiarities of manner and dispositions*' by comparing the 'human structure' with that of 'quadrupeds, birds, fishes, and insects'.[35] In other words, how far is our personality determined by our bodies? *Frankenstein* offers a dramatic response to such discussions concerning the links between a person's body and their character: is it fair to judge the Creature by his hideous outward appearance? In the novel, characters assume that the Creature is evil because of the way that he looks.

When they were in Geneva, Mary Shelley records reading 'Buffons Hist. Nat.' in June and July 1817. Smellie had translated Georges-Louis Leclerc, Comte de Buffon's *Histoire Naturelle*, a text which put forward theories of life and death on a grand scale.[36] Percy mentioned Buffon's 'sublime but gloomy theory', in a letter to his friend Thomas Love Peacock dated 22 July 1816, that the earth would be transformed to ice at some point in the future.[37] In *Frankenstein*, even when Victor is fed up with natural philosophy, he continues to enjoy reading Buffon: 'although I still read Pliny and Buffon with delight, authors in my estimation, of nearly equal interest and utility'.[38] Percy was reading 'Pliny's letters' at the end of July 1816.[39] Clearly the Shelleys read widely in areas that overlap with *Frankenstein* and the fruits of their reading can be seen in the novel.

Mary's father, William Godwin, had friendships with a number of men of science. His diary shows how many physicians, anatomists, natural historians and chemists he knew and saw regularly, and whose lectures he attended. In 1785 he anonymously translated a report written to debunk animal magnetism (*see Chapter 1*).[40] His 1799 novel *St Leon* features an alchemist who possesses the elixir of life, and his *Lives of the Necromancers* was written to expose 'the credulity of the human mind' that could believe in such ideas.[41] Similarly Victor Frankenstein is drawn to alchemy when he is young, but later rejects it in favour of the miracles performed by 'modern' chemists.[42]

The young Frankenstein chances upon a volume of Cornelius Agrippa's works and is soon hooked, but when he communicates his interest to his father, Agrippa's work is dismissed as 'sad trash'. With hindsight, Frankenstein speculates on whether 'the train of [his] ideas' would ever 'have received the fatal impulse that led to my ruin' had his father taken pains to explain 'that the principles of Agrippa had been entirely exploded'. It is characteristic that he blames someone else for the events that unfold. He misguidedly continues in his alchemical studies, procuring the whole of Agrippa's works and also reading Paracelsus and Albertus Magnus. The 'elixir of life' is his specific goal in these studies; it is not until he reaches the University of Ingolstadt that his studies are properly directed.[43]

When Frankenstein sees a tree 'utterly destroyed' by lightning and (in the later, 1831 edition of the novel) hears 'a man of great research in natural philosophy' explain 'a theory which he had formed on the subject of electricity and galvanism', he is astonished.[44] After this, the alchemists of whom he had become a 'disciple' are overthrown. Even though modern science displaces alchemy, Frankenstein sees in each equal and similar aspirations and ambitions. At university, Professor Waldman tells him that 'modern' chemists have also 'performed miracles'.[45]

Godwin knew personally, and attended the lectures of, the most famous chemist of the period, Sir Humphry Davy (*see Plate 4*). Percy Shelley bought Davy's *Elements of Chemical Philosophy* when it was published in 1812, and in 1820 wrote lengthy notes on Davy's *Elements of Agricultural Chemistry*.[46] On five occasions, in late October and early November 1816 when she was writing *Frankenstein*, Mary noted that she was reading what she called 'Davy's Chemistry' and the 'Introduction to Davy's Chemistry'. On one occasion she specifically noted that she was reading it with Percy.[47] This is one of many examples of the Shelleys reading (and writing) together during this period.

There are a number of echoes of Davy's 1802 *Discourse, Introductory to a Course of Chemistry* and *Frankenstein*.[48] Professor Waldman tells

Frankenstein that chemists 'have acquired new and almost unlimited powers'.[49] Davy uses the same language when he imagines how the chemist has powers which may almost be called creative, which have enabled him to modify and change the beings surrounding him, and by his experiments to interrogate nature with power, not simply as a scholar, passive and seeking only to understand her operations, but rather as a master, active with his own instruments.[50]

For Davy, and Waldman, chemists do not merely describe nature – they change it. This language of mastery and control comes from the earlier philosopher Francis Bacon. Nature is figured as female ('her operations') and the chemist is not a passive observer but a conqueror.

Percy Shelley's interest in science and medicine are well known. Mary and Percy had already collaborated on the journal they kept together and on the *History of Six Weeks' Tour*, which made use of their journal and was published in 1817, just before *Frankenstein*. The surviving manuscripts of *Frankenstein* reveal Percy's involvement. Charles E. Robinson calculates that Percy contributed 5–6,000 words to the 72,000-word final novel, around 13 per cent.[51] His additions often take the form of encouraging Mary to use more literary language. For example, we can see from the manuscript she originally wrote that Frankenstein's 'amusements were studying old books of chemistry and natural magic'.[52] But this rather prosaic phrase, 'old books of chemistry', is crossed out and replaced by Percy's more poetic 'I delighted in investigating the facts relative to the actual world; she busied herself in following the aërial creations of the poets.'[53] What is clear from an investigation of the surviving manuscript sources is that Mary had her own distinct and unique sense that chemistry was the crucial science in her novel. Four times Percy crossed out Mary's choice of 'chemist' or 'chemistry' to replace it with 'natural philosopher' or 'natural philosophy'.[54] In doing so, he rejects the specific discipline Mary alludes to and instead uses a more general term. The word 'scientist' had yet to be coined; 'Natural philosophy' was a broader and more genteel term than 'chemist', indicating the philosophical study of nature.[55] But it is clear from the original manuscript that Mary intended Frankenstein to be a chemist.

Despite Percy's revisions, the Victor Frankenstein of the final 1818 novel is still, primarily, a chemist. He is not a medical doctor, despite popular misconception. At university he studies natural philosophy and chemistry. Professor Waldman inspires him to prefer chemistry when he tells him that 'Chemistry is that branch of natural philosophy in which the greatest improvements have been made and may be made.' After Waldman's speech, Frankenstein is determined to excel in this discipline specifically: 'From this day natural philosophy, and particularly chemistry, in the most comprehensive sense of that term, became nearly my sole occupation.' He succeeds to such an extent that 'at the end of two years, I made some discoveries in the improvement of some chemical instruments'.[56] When he prepares to create the Creature's companion, Frankenstein specifies that he uses chemical instruments for the task.[57]

A play performed in London in 1823 based on Mary Shelley's *Frankenstein* made the novel and its characters famous. But the play version had a number of important differences from the novel. In *Presumption; or, the Fate of Frankenstein* the Creature never speaks, although he is, as in the novel, unnamed. The actor Thomas Potter Cooke was well known for his portrayal of monsters.[58] This production created a new character, Victor's laboratory assistant Fritz. The character would appear again in film versions of the novel, sometimes named Igor. In this first play, Frankenstein is described as a 'studious chemist', though he is suspected of being an alchemist.[59] Later in 1823, the title of a burlesque version of this play confirms his new identity as a medical man: *Humgumption; or, Dr. Frankenstein and the Hobgoblin of Hoxton*. It was here that 'Dr Frankenstein' was born.

A NOTE ON THIS BOOK

In this book I have decided to call the unnamed being to whom Frankenstein gives life 'the Creature', following other recent critics of the novel. This term seems so much more positive than 'Monster', and in any case, it is Frankenstein who labels the Creature a 'monster', and he is undoubtedly biased. The Creature perhaps is monstrous, though, in a number of ways. He transgresses the limits of the individual body by being made up of many bodies, both animal and human: the 'dissecting room and the slaughter-house' furnished his materials.[60] He is both living and dead, a living being made of corpses. He is also, it seems, hideously ugly and gigantic. Rather than thinking of him as subhuman, though, it is possible to think of him as posthuman. He can be seen as an improvement upon humans: more agile, vegetarian, able to survive in the Arctic wastes and South American heat more comfortably.

The narrative structure of the novel means that no single narrator is given preference. The novel starts with a series of letters from Captain Walton to his sister (who has the same initials as Mary Wollstonecraft Shelley) while on a dangerous expedition to the Arctic. When he meets Frankenstein, we hear his story told to Walton, and within that the Creature's story. The only time these frames are breached is when Walton comes face-to-face with the Creature after Frankenstein has died. Frankenstein tells Walton that he created a living being, though he refuses to tell him exactly how. The Creature – spurned and rejected by everyone he encounters – proceeds to murder Frankenstein's younger brother William (who has the same first name as Mary's then-living child), his friend Henry Clerval, and his fiancée

Elizabeth Lavenza in revenge. The second and third murders come after Frankenstein destroys a second being he had agreed to create as a companion for the friendless and unloved Creature.

I have chosen to focus on the first printed edition of *Frankenstein*, published in 1818, because this version contains more science. By 1831 the novel had become more overtly Gothic. The book focuses on the contemporaneous scientific explorations of life and death that influenced Mary Shelley's novel, rather than considering life and death as Gothic themes. Thus this book deals not with ghosts, apparitions or necromancy but with the scientific explanations for life and death given at the time. I do not mean to suggest that Mary Shelley thought solely of scientific and medical ideas when she wrote the novel – in fact, I am sure that is not true – but this book explores the many current scientific and medical theories and discoveries that potentially contribute so much to the novel. Perhaps because there was such an interest in these themes, many others wrote poetry and prose about life and death in this period. But Mary Shelley's book has surely been the most influential.

ONE
LIFE AND DEATH IN ROMANTIC LITERATURE

After going late to bed on a damp and dreary night in June 1816, Mary Wollstonecraft Godwin had a vision. As she lay there in the dark she imagined a 'pale student of unhallowed arts' had constructed something that looked like a man. Terrified, she watched as the creation began to move and live.[1] The idea that anyone would attempt such a thing appalled her. Did God alone have the right to give life? Could it even be done? She had spent the evening listening to intense discussions on 'the nature of the principle of life' between Percy Shelley, Lord Byron and John Polidori. Apparently bringing people back from the dead was more than just a fantastic proposition.

Such discussions were in the air in Britain at this time. The issue of what life was – and how to distinguish it from dead or inorganic matter – had never been more ambiguous or more worrying. Mary realized that what had terrified her would terrify others. She had decided upon the theme for her ghost story and was not the only author in this period to realize the dramatic potential of scientific experiments on the dead.

FIG. 2 *Prometheus Bound*, Christian Schussele (1824–1879), charcoal drawing.

Literature of the Romantic period was fascinated by scientific theories of life and death, and of in-between states. Authors knew about the latest scientific and medical ideas, and their writings investigated the moral consequences of these new theories. Characters in the Romantic period come back from the dead or challenge the notion of what it means to be human. Like Mary Shelley, poets and novelists turned to the imagination to try to make sense of the science of life and death.

THE ETHICS OF SCIENTIFIC EXPERIMENTATION

Early one morning in the Warrington Academy, the theologian and natural philosopher Joseph Priestley came downstairs to begin scientific experiments on a mouse that had been trapped for this purpose. In the bars of the mouse's cage was a twisted piece of paper. The note contained a poem, apparently written by the mouse itself, begging Priestley to release him from captivity. The mouse pleaded for his life, and as a result Priestley let him go. The Unitarian poet Anna Letitia Barbauld, whose father taught at the Warrington Academy with Priestley, had crept down during the night to place the poem in the bars of the cage. Her poem 'A Mouse's Petition' shows, even before *Frankenstein*, a female author trying to establish what the 'natural rights' were that all living beings possessed.

Just as the Creature in *Frankenstein* argues for his right to have a companion, Barbauld's poem gives voice to the subject of an experiment arguing for its right to live a happy, free life of its own choosing. The mouse declares that it is undeserving of such a cruel fate and charges Priestley with a duty of care. What right does Priestley have to play God? He should be friendly and hospitable, not a torturer or an executioner. The 'well taught philosophic mind', the mouse reproaches Priestley, should show compassion to everyone.[2] The

FIG. 3 *The Mouse's Petition,* William Ward after James Ward, 1805.

mouse's argument resonates with the fact that the Creature had no say in Frankenstein's hideous experiment. Likewise, the Creature demands of his maker: 'How dare you sport thus with life?'[3]

Many mice died in Priestley's experiments with gases in the early 1770s. As a result of these experiments Priestley isolated oxygen, though he never called it by this name. The mouse in Barbauld's poem argues that 'vital air' is a 'gift of heaven', given to all regardless of their wealth, social or species status.[4] It tells us that oxygen is not something that should be controlled by others. The mouse raises the same concern that people were raising at this time: what are the inalienable rights of humans? Often it was specifically men's rights being discussed, rather than, say, human rights. For example, the

French Declaration of the Rights of Man and of the Citizen declared that 'resistance to oppression' was one such inalienable right that all men should possess.

Writers were intrigued by what differentiated – or connected – living beings. Samuel Taylor Coleridge's poem 'The Eolian Harp', like Barbauld's mouse, also tried to find connections – rather than differences – between all living life forms. There is, he writes, 'one life within us and abroad', connecting humans to animals and plants.[5] The mouse tells Priestley in Barbauld's poem that he should feel equally for 'all that lives', whether this be mice or men.[6] What are the responsibilities of the scientist or medical practitioner? These texts give voice to early concerns regarding scientific explorations and discoveries.

The question of animal rights emerged in the Romantic period, as did the rights of women, slaves and children. Supposed biological differences have been used throughout history to justify oppression and subjection, but Barbauld's poem features a potential victim speaking back to a captor and demanding that he review his decision. The mouse questions whether there is life after death, and, if there is not, he asks what right Priestley has to decide who lives and who dies. This question is a poignant one for *Frankenstein*. The mouse ends with what sounds like a threat to Priestley. If you are kind to others in this life, that may help you in the next.

Barbauld returns to the subject of what life is in another poem, titled 'Life'. She confesses that while she still is yet to understand exactly what life is, the only certainty is that she will eventually part company with it. In other words, she knows that she will die. She describes the dead body as 'valueless', a 'clod', and wonders 'what then remains of me' after death?[7] The Creature in *Frankenstein* is made up of the bodies that Barbauld here describes as without value.

WHAT IS LIFE?

Samuel Taylor Coleridge was a frequent visitor to William Godwin's house. Mary remembers hearing him recite his scary and hypnotic poem *The Rime of the Ancient Mariner* while she hid behind a sofa as a child. Her journals record Percy Shelley reading the poem aloud in September and October 1814, and she refers to the poem a number of times in *Frankenstein*.[8] In Captain Walton's second letter to his sister, he writes that he, like Coleridge's Mariner, is exploring the 'land of mist and snow', though the Mariner has been in the Antarctic and Walton is in the Arctic. Unlike the Mariner, Walton declares, he shall 'kill no albatross'. Victor Frankenstein quotes Coleridge's Mariner when he abandons his Creature after he has given him life. Frankenstein spends the night pacing sleeplessly in a turmoil of horror and guilt. Like the Mariner, Frankenstein does not need to turn his head to know that the 'frightful fiend' is behind him. When he contemplates his wedding to his cousin, Elizabeth, Frankenstein feels the 'deadly weight' of his promise to create another Creature 'hanging round his neck' and 'bowing [him] to the ground'.[9]

When the Ancient Mariner kills the albatross for no reason, this act of unmotivated violence sets in motion the series of events that will lead to his return as 'a grey-beard loon', terrorizing people with his surreal and frightening story. The poem features extreme conditions for life as well as in-between states of the living dead. The Mariner tells of the nightmare figure called 'Life-in-Death' who wins the lives of the crew at dice. As they die, each crew member curses the Mariner, and he experiences severe loneliness until able to bless the water snakes and break the curse. A horror similar to that of *Frankenstein* is recalled, as the corpses of the crew, now 'ghastly', rise again, using their 'limbs like lifeless tools' to get him back to shore.[10]

Coleridge's poem shares its mood with Shelley's *Frankenstein*: both protagonists commit an atrocious act for which they pay dearly. Both poem and novel are suffused with foreboding and dread. Both protagonists tell their stories to someone who is changed by the experience, though the tale offers the storyteller little solace. The Mariner, like Frankenstein, knows that he has done something wrong, the consequences of which extend far beyond his own person and also destroy those nearest to him. Unlike 'A Mouse's Petition', in Coleridge's *Rime* we hear from the perpetrator rather than the victim. In *Frankenstein* we hear from both and this adds to the power of the novel. The 2011 National Theatre production that alternated the roles of Frankenstein and the Creature between actors Benedict Cumberbatch and Jonny Lee Miller played upon this idea of duality.[11]

Throughout his life, Coleridge analysed his dreams and nightmares, and the effects of opium on his mind. He was an early admirer of Priestley, was initially seduced by John Brown's theory that in life we are allotted a certain amount of 'excitability' (*see Chapter 5*), and he admired J.F. Blumenbach's idea that there was some formative power within us that is responsible for generation (*see Chapter 4*). He became a close friend of the chemist Humphry Davy and considered Davy's research to be leading us 'to the door of the temple of the mysterious god of Life'.[12]

Coleridge took the physiological ideas of life that he gained from Brown, John Hunter and Blumenbach, and applied them to the imagination. Using scientific language, Coleridge describes the imagination as '*essentially* vital' and 'the living power'.[13] He writes that 'The *rules* of the IMAGINATION are themselves the very powers of growth and production.'[14] This was only one example of Coleridge's fascination with science and medicine. Psychology was only just emerging as a scientific discipline at this time but he was already extremely interested in this 'science of the mind'.

While many of Coleridge's scientific enthusiasms were short-lived, such as when he told Davy he would 'attack Chemistry, like a Shark',[15] his interest in the question of what life was resulted in his writing *Hints towards the Formation of a more Comprehensive Theory of Life* (1818).[16] This text took its place among the debate in the Royal College of Surgeons between John Abernethy and William Lawrence, from 1814 to 1819 (*see Chapter 4*). Coleridge stood firmly with Abernethy, who had treated him as a patient. Attending Abernethy's Hunterian Oration in 1819, Coleridge heard himself quoted, a fact which demonstrates just how seriously his scientific opinions were taken.[17]

In answer to the question, 'What is Life', Coleridge answered with his own idiosyncratic theory: 'I define life as *the principle of individuation*, or the power which unites a given *all* into a *whole* that is presupposed by all its parts.'[18] In other words, it is the power that unites parts into a unique and individual whole, living being. This is very similar to the kind of wholeness a poet can achieve through his creativity. As Coleridge writes in *Biographia Literaria*, a poet 'diffuses a tone and spirit of unity, that blends, and (as it were) fuses, each into each, by that synthetic and magical power [called] Imagination'.[19]

Again and again we can see Coleridge take up scientific ideas and apply them to literary topics. In 'Human Life, On the Denial of Immortality, A Fragment', Coleridge asks: is our physical body all there is of life? Does our physical presence (the 'sound and motion' of our bodies) really comprise 'the whole of Being'?[20] In *Frankenstein* many of the characters assume a link between the Creature's exterior and his inner self but the novel asks us to question this assumption.

Coleridge contemplates various theories of what life is. He asks, for example, whether 'Breath / Be Life itself'? In other words, is it true that breathing oxygen is really why we are alive? Will anything continue beyond our physical death into an afterlife? Even while he considers various possibilities, it is clear that Coleridge does not believe our lives are merely 'summer-gusts, of sudden birth and doom'. He refuses to believe that human life is 'purposeless, unmeant', merely the 'surplus' matter of nature's activities. Despite this, the poem ends equivocally: 'Thy being's being is contradiction.'[21] Ultimately, then, for Coleridge life is unknowable, ambiguous and even a paradox.

THE DANGER OF ACQUIRING (ILLICIT) KNOWLEDGE

When Percy and Mary met up with Byron at Sécheron in Geneva, Switzerland, on 27 May 1816, Mary's stepsister Claire Clairmont and Byron were already lovers. Claire was pregnant with his child either before or soon after their renewed acquaintance in Switzerland. The fate of their child, Allegra, who died aged 5 in 1822, is yet another poignant reminder of how tenuous life was in those times.

Byron and Shelley instantly bonded, and before long the Shelleys and Clairmont had become neighbours of Byron and Polidori. The weather was appalling, to the extent that 1816 became known as 'the year without a summer'. Many years later, it was understood that the eruption of a volcano, Mount Tambora in Indonesia, had been the cause. Perhaps for this reason weather is much mentioned in *Frankenstein*. Byron was also inspired by the weather conditions and the discussions that took place that summer. His poem 'Darkness' imagines the sun being extinguished and the consequent chaos and horror that would ensue.

In the first canto of *Don Juan*, Byron displays his scientific and medical knowledge. Edward Jenner's vaccination against cowpox and Davy's miners' safety lamp should be thought of as representative of the age, he argues. But there are also some more worrying developments afoot:

> And galvanism has set some corpses grinning,
> But has not answer'd like the apparatus
>   Of the Humane Society's beginning
> By which men are unsuffocated gratis.[22]

Byron claims that galvanism has not proved as successful a remedy for suffocation as the methods of the Humane Society (*see Introduction*). Such ideas appear in Mary Shelley's 1831 Introduction to *Frankenstein*: 'Perhaps a corpse could be reanimated; galvanism had given token of such things.'[23]

In the dramatic poem *Manfred* (*see Plate 8*), written in the autumn of 1816, Byron asks us whether scientific minds should be held accountable for their actions. Is it not an essential characteristic of the man of science to push boundaries and challenge convention? Manfred is an alchemist figure, who reveals his supernatural powers by raising spirits. In fact, his education sounds a great deal like Victor Frankenstein's:

> And then I dived,
> In my lone wanderings, to the caves of death,
> Searching its cause in its effect; and drew
> From wither'd bones, and skulls, and heap'd up dust,
> Conclusions most forbidden.[24]

Both men pursue their investigations secretly and alone. Both explore the state of death in order to discover the principle of life. Both experiment on the corpses of the dead. Both also, arguably, discover knowledge that should not be possessed by mere mortals.

Manfred is arrogant like Frankenstein and he infuriates the spirits he raises. They are outraged to be summoned by a puny, insignificant mortal being. Manfred thinks himself superior to the rest of humanity and tells the lowly chamois hunter, who has just saved him from attempted suicide, 'I am not of thine order.' In this statement he refers both to his aristocratic lineage (Manfred is a lord who lives in a castle) and to his belief that he is not like other men. He speaks of other people disparagingly as 'breathing flesh' and 'creatures of clay' and claims that the powers he possesses alienate him from them.[25] Frankenstein is strikingly like Manfred; equally monomaniacal, he also puts his own interests above the interests of others.

'Child of clay' is a favourite phrase of Byron's, which he often uses to describe human life. Since ancient times, myths across many different cultures and religions have imagined that humankind was created from clay. According to Manfred, humans are 'Half dust, half deity': they have a 'mix'd essence'.[26] The clay or dust is the physical, bodily part of us with its base desires. Our souls are the godlike part of us that partake of heaven. Whether the Creature has a soul is a vital unanswered question in *Frankenstein*. He certainly does possess an emotional and intellectual life beyond the purely physical.

The full title of Mary Shelley's novel, *Frankenstein; or, the Modern Prometheus*, alludes to the Greek myth in which Prometheus gives fire to man.[27] Fire is often a symbol of the vital spark that animates living beings. The novel's subtitle may also allude to other aspects of Prometheus's character, drawn out by Byron's poem 'Prometheus', which was published in July 1816, and Percy Shelley's later dramatic poem *Prometheus Unbound*. It is clear that the figure appealed to both poets. They both seemed to identify with this figure because he is a suffering rebel, punished by Zeus for stealing fire (*see Figure 2*). Byron's Prometheus is unrepentant; he is 'a symbol and a sign' of what humanity is and can be in its heroism.[28]

Like Manfred, Prometheus shares many characteristics with Victor Frankenstein. He is alone and his fate is to continue forever alone. He meant well when he gave humans fire; in Byron's poem, his 'Godlike crime was to be kind'. His efforts were meant to render 'The sum of human wretchedness' less than it had been previously.[29] When Frankenstein first comes up with the idea of creating a living being he considers how 'Godlike' this would make him: 'A new species would bless me as its creator and source.'[30] While this is blatantly narcissistic and egotistical, it also offers a potentially beneficent reason for Victor's experiment. The early, and preventable, death of Victor's mother from scarlet fever may well have motivated his thinking. What if we could bring back our lost loved ones from death?

Byron and Shelley admired and emulated Prometheus; they felt themselves equally to be exiles and outcasts, made to leave England because of a sanctimonious moral conduct they had transgressed with their sexual scandals. By aligning Victor Frankenstein with such a figure, Mary Shelley seems to be criticizing her lover and friend. Even Frankenstein admits the foolishness of his scientific aspirations when he tells Captain Walton:

> Learn from me, if not by my precepts, at least by my example, how dangerous is the acquirement of knowledge, and how much happier that man is who believes his native town to be the world, than he who aspires to become greater than his nature will allow.[31]

Yet, Frankenstein's warning here sounds a little hollow. Prometheus, Manfred, Frankenstein and Walton are all overreachers. Perhaps science requires boundaries to be pushed and even transgressed. Stem-cell research today has caused similar concerns. Does science demand the acquisition of dangerous knowledge? The fact that Frankenstein is a flawed character, and an unreliable narrator for this reason, adds to the novel's interest.

MONSTROUS BEINGS: AUTOMATONS, SLEEPWALKERS & VAMPIRES

John Polidori's short and unhappy life has had great impact on literature and culture. He took his medical degree very young, aged 19, at the University of Edinburgh. It was a fashionable place to study medicine, but as a consequence was suffering from problems of overcrowding. Due to a series of nepotistic appointments, the teaching was not of the highest quality. Medicine had been his father's choice for him, rather than his own, and Polidori was not happy. Being too young to apply for a licence to practise medicine, he went to Geneva with Byron, then 28, as his travelling companion and personal physician. While it is likely that they were lovers, Byron quickly became bored and the group seemed to have ridiculed Polidori's literary aspirations.[32] The whole experience was horribly disappointing. Fired by the end of the summer, Polidori walked to Italy and travelled there for a short time. When he returned to England he tried and failed to set up a medical practice. He committed suicide in 1821 aged 25.

Polidori's knowledge aided the late-night conversations that inspired *Frankenstein*. Indeed, his diary records that on 15 June 'Shelley and I had a conversation about principles, – whether man was to be thought merely an instrument.'[33] During his medical education, Polidori would have been familiar with the practice of grave-robbing, the typical means by which medical students obtained corpses for anatomy (*see Chapter 5*). He would also have known about contemporary theories of life, including the current debate between Abernethy and Lawrence on this topic. Half-Italian, Polidori described the English at the university as 'automatons' for their lack of enthusiasm and passion in a letter to his father; he seems to have meant that they lacked compassion and emotion.[34]

Polidori's contribution to the ghost-story competition was titled *Ernestus Berchtold; Or, the Modern Oedipus*, and it was published in

1819. Byron's effort was never completed, though the fragment he wrote was later published. However, Polidori took Byron's outline for his tale and breathed new life into it. The resulting story, 'The Vampyre', was published in the *Monthly Magazine* on April Fool's Day in 1819, with Byron's name given as the author, though both Byron and Polidori quickly asserted that Polidori was the tale's true author. This is the first example of the aristocratic vampire with whom we are now familiar, from Bram Stoker's *Dracula* to Anne Rice's *Interview with the Vampire*. In *Frankenstein*, Victor calls the Creature 'my own vampire, my own spirit let loose from the grave, and forced to destroy all that was dear to me'.[35]

Byron had mentioned vampires in *The Giaour*, cursed to 'drain the stream of life' from female members of their family in order to 'feed thy livid living corse'.[36] On 18 June 1816 Byron recited Coleridge's 'Christabel' and the potentially vampiric central character scared Percy Shelley in particular. According to Polidori's diary, he ran out of the room in terror.[37] Vampyrism and science overlapped more than we may now assume. The first recorded successful blood transfusion from human to human was soon to be achieved when the obstetrician James Blondell transfused blood from a husband to wife in 1818. Such movement of living matter from one person to another was fraught with unknown consequences and it was generally thought that the characteristics and behaviour of the person would be transferred along with their blood.

Like the Creature in *Frankenstein*, vampires exist in a state between life and death. They are able to exercise their will over their victims before sucking the lifeblood from them. Polidori clearly felt that Byron used and abused people in just this way, albeit metaphorically. Fed up with being the butt of the group's jokes, Polidori took his revenge by calling his vampire Lord Ruthven, a name Caroline Lamb had given her fictionalized version of Byron in the novel *Glenarvon* (1816). Polidori's vampire is cold, unfeeling and heartless. His 'dead grey eye' and bloodless, 'colourless cheek' are remarked upon at the beginning of the story.[38]

In 'The Vampyre', the aristocratic Ruthven, whose 'affairs were embarrassed', travels abroad with the young, naive Aubrey, an orphan whose head has been turned by reading too much poetry in his youth.[39] Just before he is killed, Ruthven makes Aubrey swear that he will tell no one of his death. Thus, when Ruthven comes back to life, living among London society once again, Aubrey is unable to reveal the truth. The relationship between Ruthven and Aubrey mirrors that of Byron and Polidori with its unequal distribution of power, the younger in thrall to the elder.

Polidori clearly shared the Romantics' fascination with alternate states of being. His final-year undergraduate dissertation was on the topic of oneirodynia, a term used to describe disturbed sleep, including nightmares and somnambulism. Sleep was considered to be a state of suspended animation. Drug-induced or mesmeric trance, fainting and coma were similar states when the body seemed to function automatically or mechanically.

Then, as now, there was a question over whether sleepwalkers should be held responsible for their actions. William Wordsworth's poems 'The Somnambulist' and 'The Brothers' present local Lake District legends concerning the potentially disastrous consequences of sleepwalking. In the first poem, Lady Emma falls into the rapids at Aira Force and dies upon being startled awake by her lover. In the second, we hear of the younger brother James sleepwalking to his death over a precipice. Charles Brockden Brown's *Edgar Huntly; Or, Memoirs of a Sleep-Walker* (1799) also explores the consequences of this disorder. Polidori notes in his diary that somnambulism was a topic of his conversation with Genevan men of science in the summer of 1816.[40] Sleep does not always offer repose in *Frankenstein*; while there is no mention of sleepwalking, Victor suffers from nightmares at various points in the novel. After the murder of Clerval, beset by fever, he raves in his sleep, frightening others with his entreaties, wild gestures and bitter cries.

As the Industrial Revolution progressed, distinctions became blurred between machines and humans. The Luddite movement began in the Romantic period as a protest against the use of machines to replace human labour in factories. At the same time, mechanical beings were created that looked and moved as though they were human and this inevitably provoked questions about the possibility of recreating human life. Erasmus Darwin, Charles's grandfather, created a speaking wooden head that could utter words such as 'mamma' and 'papa' so authentically as to deceive those who heard it into thinking it was a human voice.[41] 'The Turk' appears to have been an automaton chess player, created in 1770 by Johann Wolfgang Ritter von Kempelen de Pázmánd (*see Plate 10*). It was eventually proved to be an elaborate hoax; there was, in fact, a human operator hidden inside. Automata feature in E.T.A. Hoffman's short stories 'The Automaton' (1814) and 'The Sandman' (1816). Polidori described Byron as being like an automaton; he found him cold and unfeeling during their time in Geneva. When he accidentally struck Byron on the knee with an oar while rowing, Polidori said he was glad to discover that Byron could feel pain.[42]

Hoffman was fascinated by uncanny versions of human life, such as animal magnetism, made popular by Franz Friedrich Anton Mesmer, which Hoffman wrote about in 'Der Magnetiseur'. Likewise, Percy Shelley was convinced that mesmerism worked; his poem 'The Magnetic Lady to her Patient' (1822) describes abdicating power

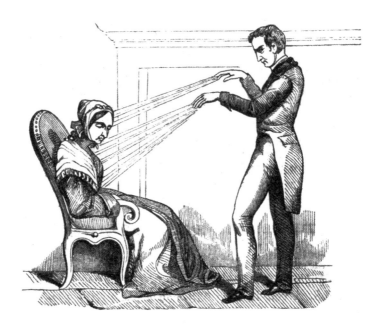

FIG. 4 A practitioner of mesmerism using animal magnetism. Wood engraving, late eighteenth or early nineteenth century.

over oneself to another. Mesmerism, sleepwalking and other states of suspended animation were thought of as episodes where humans became like automata, their will subsumed by another (in the case of mesmerism) or unknowable forces of the subconscious (in the case of somnambulism). They became instruments in the hands of others. Though the Creature might be thought of as an automaton, possibly created without a soul, many readers consider Frankenstein to be the real monster, lacking human feelings of compassion towards his creation.

One afternoon in 1797, after taking an 'anodyne', which was almost certainly opium, because he felt unwell, Coleridge claimed that the whole of the poem *Kubla Khan* came to him in a kind of dream.[43] He immediately began writing the lines that we now have. This task was rudely interrupted, he said, when a 'Person from Porlock' knocked on his door, destroying his concentration. In her 1831 Introduction,

Mary Shelley describes the moment of her inspiration for *Frankenstein* in the same way: 'When I placed my head on my pillow I did not sleep nor could I be said to think. My imagination, unbidden, possessed and guided me, gifting the successive images that arose in my mind with a vividness far beyond the usual bounds of reverie.'[44] This is not a dream but more like a vision. Though she is awake, she is not in control: her imagination is 'possessed'. Mary represents herself as existing in an in-between state, in which will and volition are powerless. She is inspired by something outside herself, resembling the kind of passive automaton in which the Romantics were so interested.

THE MUTABILITY OF LIFE

In 1815 Mary visited Oxford University and saw Percy's student rooms, where he had experimented with an electrical machine, air pump, galvanic trough and solar microscope (*see Plate 15*). He had even intended at this point in his life to become a doctor. In 1811 he wrote to a friend, 'I still remain firm in my resolve to study surgery – you will see that I shall.'[45] After he was expelled from Oxford, following the publication of the *Necessity of Atheism*, Percy went to stay with his cousins, John and Charles Grove, in Lincoln's Inn Fields, London, where the Royal College of Surgeons was located. John had already qualified as a surgeon but Charles was training to become one and Percy attended a course of lectures and walked the wards of St Bartholomew's Hospital with him.[46]

Percy was particularly intrigued by the question of what life was. He wrote an essay in late 1819, now given the title 'On Life', in which he muses on the way that 'The mist of familiarity obscures from us the wonder of our being.' Life, he writes, is more of a mystery than the creation and destruction of the planets, or of empires, religious and political systems on this earth. Percy expresses his admiration at life's 'transient modifications' but the fact of life itself is, he writes, the 'great miracle'.[47]

For Percy, life was characterized by change. A number of his poems, particularly those written in 1816, discuss life as a series of transformations. They meditate on the transformations that take place from life to death and from death to life. One poem, 'Mutability', is quoted twice in *Frankenstein*, demonstrating Mary's sense of affinity between the two texts. In this poem, Percy compares humans to clouds that move 'restlessly' in the sky, radiant for a time but, eventually, lost to sight as night comes.[48]

In 'Mutability', Percy imagines humans as aeolian lyres, a metaphor also used by Coleridge in his poem of that name, an instrument that sat on the windowsill or hung from a tree branch and was played by the wind. By means of this metaphor, Percy imagines that humans are not in control of their emotions, that they respond differently each time to external forces, and that no response is the same as the last. Even when we rest, he writes, dreams can 'poison' our sleep. Then, when we wake up, a chance thought can ruin the rest of our day. In this poem, humans seem at the whim of random forces. Exerting no power of their own, they are directionless and without purpose. His grand conclusion seems to be that none of this matters: 'It is all the same!' The only thing that will survive of us is change: 'Nought may endure but Mutability.'[49]

Percy Shelley was fascinated by death. If his poems are to be believed, when he was young he lay in coffins and chased ghosts. In *Alastor; Or, the Spirit of Solitude*, written at the end of 1815, the poet-figure hopes to achieve some answers to the 'obstinate' questions of 'what we are'.[50] Percy wonders whether death is the same changeful state as life or the point at which there is no more change. In 'Hymn to Intellectual Beauty', conceived on a boating trip with Byron in

June 1816, he again presents life, or 'what we feel and what we see', as uncertain and random.[51]

Percy's poem 'Mont Blanc' was inspired by a visit to Chamonix, the influence of which can also be seen in the sublime alpine scenes in *Frankenstein*. As this and other poems make clear, he thinks that existence is made up of a cycle of life and death: 'All things that move and breathe with toil and sound / Are born and die, revolve, subside and swell.'[52] He imagines the world in constant transformation as life is created, destroyed and changed into new forms. His views on life influenced his vegetarianism. Percy's 1813 *A Vindication of Natural Diet* ends with 'NEVER TAKE ANY SUBSTANCE INTO THE STOMACH THAT ONCE HAD LIFE.'[53] The Creature in *Frankenstein* is also a vegetarian.

It is difficult to believe that life is a kind of cycle, as Percy did, when an individual dies. In the poem written on the occasion of Keats's death, *Adonais*, Percy moves through a series of sometimes conflicting ideas of what death is. The changes undergone during life are all for the worse: '*We* decay / Like corpses in a charnel'.[54] Likewise, in her essay 'On Ghosts', Mary Shelley described living mortals as 'walking corpses' when compared to the dead.[55] In Percy's poem there is some optimism though. Keats, now at peace, 'is made one with Nature'.[56]

Percy's meditations on the nature of life and death were informed by the scientific and medical knowledge he gleaned from many sources. The French anatomist Marie François Xavier Bichat famously declared that life was that which resisted death, suggesting that life is a process of decay. The doctor Sir Thomas Charles Morgan (husband of the novelist Sydney Owenson) questioned whether we remain the same person during our lifetime because of the changes that take place in our bodies. Morgan argues that due to changes including the loss and regrowth of skin, nails and hair, we are a new person every forty days.[57] These ideas are embodied in the Creature, a walking, talking figure, created from the charnel house and the graveyard, a corpse who is yet alive.

LIFE AS SUFFERING

As an apprentice to an apothecary, aged 14, John Keats learned to prepare the leeches that were used to suck blood from patients. He prepared medicines, and perhaps also dressed wounds and pulled teeth. He entered Guy's Hospital on 1 October 1815, and on 25 July 1816 qualified for his apothecary licence. During his time at Guy's he would have dissected corpses brought to the hospital by grave robbers and learned about the theory and practice of medicine. Keats also had much personal experience of disease and death, including nursing his brother Tom through the final stages of tuberculosis. This was the disease which would also kill Keats, at the age of 25, on 23 February 1821. Percy Shelley met Keats a few times, though there's no evidence that Mary did. Percy urged Keats to come to Italy when he heard that he was ill.

Keats was fascinated by states beyond ordinary consciousness. At the end of 'Ode to a Nightingale' he asks, finally, whether the experience of listening to the nightingale was 'a vision, or a waking dream?' The poem concludes with the question 'Do I wake or sleep?' A trance-like state is induced from the opening lines where the poet describes how he feels a 'drowsy numbness' as though he had taken some poison, such as hemlock, or drug, like opium. The poet wishes that he could, like the nightingale, 'Fade far away, dissolve, and quite forget' the world of human mortality: 'Where youth grows pale, and spectre-thin, and dies'.[58] He imagines that poetry can do this, just as the urn in his 'Grecian Urn' ode can. These forms can achieve a timeless existence beyond the horror of physical suffering which typifies human life.

In 'This Living Hand', Keats imagines his physical hand, 'now warm and capable', reaching out after death to grasp the reader from the cold and icy tomb. The poem describes the physical changes that he will go through. He imagines being able to 'haunt' the reader, even from the beyond the grave, such that the reader would wish their own heart bled dry to once again fill the poet's veins with blood. The result is a chilling poem, with a kind of blood transfusion providing the metaphor. It ends with the poet holding out his living hand towards the reader: 'see – here it is – / I hold it towards you.'[59] The dashes serve to offer visual signs of the movement of the hand reaching out over the centuries towards us today. For Keats, human life was one of suffering and pain; poetry offered the possibility of a different kind of existence.

Victor Frankenstein's life in the novel similarly has little happiness in it. By the time that Captain Walton meets him, he has lost almost everyone he loved in his life and has long given way to dark, suicidal thoughts. Life has ceased to have meaning for him. The Creature too has submitted to feelings of hatred and revenge. It is possible that these characters display features of Mary Shelley's and others' lives at this time, such as the loss of her first baby and the humiliation of Polidori at the hands of the unfeeling Lord Byron. Life was clearly tenuous and vulnerable in this period. Of all the texts considered in this chapter, Mary Shelley's *Frankenstein* most clearly articulates the questions surrounding life and death.

TWO
VITAL AIR

We know now that oxygen is the vital component of air that we need to live. But this fact had been discovered only a few decades before the first publication of Mary Shelley's novel in 1818. When Victor Frankenstein is first found by Captain Walton's crew, he is described as being in a 'wretched' condition. His body is frozen and emaciated. The crew try to take him down to a cabin but 'as soon as he had quitted the fresh air he fainted'. They immediately bring him back up to the deck and attempt to restore him to 'animation', or, in other words, to bring him back to life. The fresh air does some good but they also employ the Royal Humane Society's method of 'rubbing him with brandy'.[1] The episode shows that Mary knew what to do in an emergency of this kind, that she was aware of the most recent medical advice of the time, and that she recognized how vital air was to life.

In the novel, air is also recognized as being important for health. Romantic poets praised the good, clean air of the country as opposed to the pollution of the new cities. Victor's younger brother Ernest, who had been sickly as a child but is now healthy, is described as being 'continually in the open air, climbing the hills, or rowing on the lake'. These activities symbolize his newly acquired strength and vigour. When the weather permitted, they were exactly the kinds of activities enjoyed by the Shelleys, Lord Byron, John Polidori and Claire Clairmont in that summer of 1816. In the novel, when Henry Clerval comes to Ingolstadt to look after Victor, they take a fortnight's walking tour in the city's environs. The latter's health begins to improve in part because of the wholesome air: 'my health and spirits had long been restored, and they gained additional strength from the salubrious air I breathed.'[2] Mary Shelley realizes that fresh air can return people to health and happiness.

Air is described as 'free' by Felix De Lacey, alluding to its identity as something to which all living creatures have a right. It is also described as 'fresh', 'pure' and 'soft' at other moments in the novel.[3] The Creature feels the benefits of air upon his mood and sense of well-being: 'the fresh air and bright sun seldom failed to restore me to some degree of composure'.[4] In the novel, air revives, restores and refreshes. It brings people back to life metaphorically and literally.

THE AIR-PUMP

Victor Frankenstein tells us that his 'utmost wonder was engaged by some experiments on an air-pump' shown to him by a gentleman when he was young.[5] Perhaps the most famous illustration of the air pump is given in Joseph Wright of Derby's 1768 painting *An Experiment on a Bird in the Air Pump* (*see Plate 12*). The painting invokes a dramatic scene. An itinerant lecturer has been employed by a wealthy merchant to educate his family in scientific matters. The lecturer is the only one who looks out to us, the audience, directly. All others in the scene are looking in different directions. He seems to ask us how we feel about what is happening here. The only other living creature that – it might be argued – seems to be looking directly at us is the bird. The exotic white cockatoo has its head turned our way, one wing out, as it lies drooping in the glass bowl. We might be reminded of the mouse in Anna Barbauld's poem, which challenged us and the experimenter, Joseph Priestley, to save him (*see Chapter 1*). The experiment shown in the painting proves that air is essential to life. The air has been removed from the bowl by the lecturer, creating a vacuum. Without the air it needs to breathe, the bird flutters to the bottom of the glass jar.

What we hope – and expect – will happen next is that the lecturer will reintroduce the air and the bird will fly up again, revived, thus demonstrating the importance of air. But the moment that Wright chooses to paint is the moment before the bird is saved. At this moment we are unsure whether the bird will succumb to the loss of oxygen and die. Witnessing this experiment is an uncomfortable experience and the merchant's two daughters look the most uneasy. The youngest looks up at the bird, visibly upset, while the older daughter looks away, shielding her eyes. Their father points at the experiment, trying to make her look at what is happening.

The other figures in the scene are less easily identifiable: the boy opening the window is perhaps the lecturer's assistant. The young, expensively dressed couple are perhaps relatives of the householder. They have eyes only for each other. The young boy next to them is leaning over, perhaps trying to see how the experiment works. It is difficult to establish the sight line of the younger man sitting at the table, though it seems to be beyond the glass bowl. The older man, sitting on the other side, is looking at a jar of organic matter on the table, lost in thought. He has removed his glasses. He leans on his stick and seems to be contemplating the march of science, perhaps not positively. There are various instruments and apparatus on the table, including this glass vessel that is lit up beneath the bird and seems to contain once-living matter of some sort.

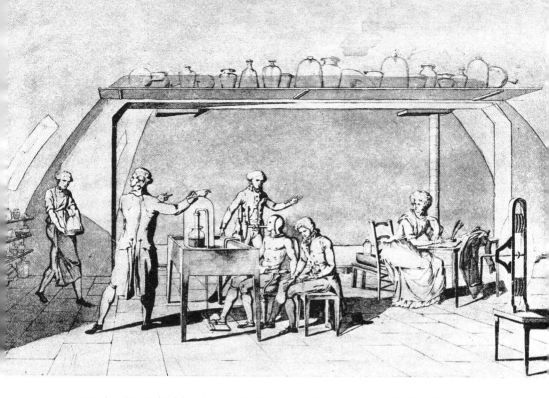

FIG. 5 Lavoisier in his laboratory conducting an experiment on the respiration of a man at work. Photogravure after M.A.P. Lavoisier, c.1850.

As a view of contemporary science, this scene is ambiguous, at best. Priestley asserted the potential power of science when he wrote that 'the English hierarchy (if there be any thing unsound in its constitution) has equal reason to tremble even at an air pump or an electrical machine'.[6] Mary Shelley played upon the fear that reviving a creature could provoke – whether involving an air pump or an electrical machine – when she imagined a creature animated from the bodies of the dead.

OXYGEN

Professor Waldman, Frankenstein's inspirational lecturer at Ingolstadt University, mentions understanding 'the nature of the air we breathe' as one of the main achievements of 'modern chemistry'.[7] Indeed, this was a relatively recent discovery. Figure 5 shows the 'father of modern chemistry', Antoine Lavoisier, performing an experiment on the respiration of a man at work. Lavoisier was guillotined in the French Revolution in 1794, but not before he had established a number of important and long-lasting changes to the discipline of chemistry.

Lavoisier was one of no fewer than three people who claimed to have discovered the element oxygen. He was, though, the person to give it the name that has persisted to this day. In fact, he introduced a whole new system of nomenclature, whereby the name given to new elements presented an idea of their character. In this case, 'oxygen' means a maker, or creator, of acids. Unfortunately, Lavoisier was incorrect in his assumption that all acids contained oxygen, but despite this the name has endured.

Lavoisier overthrew the theory, to which Priestley and others subscribed, that a substance called phlogiston was emitted during combustion. Phlogiston was described as an 'imponderable' substance because it could not be weighed or seen. Lavoisier instead believed that a fluid called 'caloric' explained heat and cooling. When Priestley discovered what we now call oxygen in 1774, he named it 'dephlogisticated air', confirming his continued belief in the existence of phlogiston. Both men, and Karl Wilhelm Scheele, a German–Swedish chemist who also claimed to have isolated oxygen first in 1771, recognized the importance of oxygen in combustion. As a result, Scheele called oxygen 'fire air'. Professor Waldman perhaps refers to Lavoisier, while recognizing the work of alchemists, when he tells Frankenstein that 'modern' chemists perhaps had an easier task than their ancient forebears: 'to give new names, and arrange in connected classifications' the facts their predecessors had discovered.[8] While chemists could not agree on a name for oxygen, they all agreed on its importance to life.

The Creature in *Frankenstein* also recognizes the importance of air to fire. When he recounts his earliest experiences, he recalls one day, when he was particularly cold, finding a fire left by some beggars. Rather in the manner of an experimentalist himself, he gradually works out what is needed to make the fire burn. He establishes that he needs dry – not wet – wood, for example. He sees that 'a gentle breeze quickly fanned it into a flame'. This observation leads him to construct 'a fan of branches, which roused the embers when they were nearly extinguished'.[9] Without realizing it, the Creature has worked out here that oxygen is needed for substances to burn. When he sets the De Lacey family's cottage on fire in revenge for their treatment of him, he notices again that the 'wind fanned the fire'.[10]

The new insight into the role of oxygen had implications for understanding how life worked too. Lavoisier's experiments on respiration showed that oxygen was breathed in, but carbon dioxide was breathed out. He hypothesized that breathing was an act of combustion. It was this operation that gave the body heat. It was, in other words, essential to life. The ability to stay warm was considered one of the key characteristics of the living body. So-called 'animal heat' was sought by those attempting to decide whether a person was living or dead. This is why warming the body figured so highly in the Royal Humane Society's recommended method for reviving people who were near death. In the 1831 introduction to *Frankenstein*, Mary Shelley imagines that a creature might be 'endued with vital warmth', showing that she understood the importance of heat for life.[11]

Priestley's work also led to our modern understanding of photosynthesis. He found that a mouse left in a sealed container would soon suffocate because the air had become vitiated. But the introduction of a plant into the container meant that the mouse lived longer. He realized that the plant replenished the oxygen needed for life. A Dutch physician, Jan Ingenhousz, established that this experiment only worked if the plant was kept in light. By these means, the importance of water and light to the growth of plants was understood. There was also a growing understanding that plants underwent a form of respiration, just as humans did. They absorb food in the same way and have a similar vascular system.

In the spring of 1820 Percy Shelley made eighteen pages of notes on these kinds of topics from Humphry Davy's 1813 book *Elements of Agricultural Chemistry*. While his notes were written two years after the first publication of *Frankenstein*, they reveal something of Percy's knowledge of the subject. Potentially, they might be evidence for what Mary Shelley knew too. Percy demonstrated that he understood what we would now call photosynthesis when he wrote 'Light necessary to the health of plants'. When listing the elements, Percy defines oxygen by its vital function – 'Oxygene (the principle of animal life)' – even though Davy had not mentioned this function in the text that he was copying from.[12] Percy found in Davy's book scientific evidence for his view that all living beings were similar in important ways. Referring to sapwood, the softer part of the tree trunk between its bark and harder wood, Percy notes 'The alburnum is the great vascular system of the plant.'[13] This is evidence of a universal life in all living beings. Comparative anatomy found equivalent vital processes in animals, humans and plants.

In one section of his notes, Percy records exactly the kinds of processes that animals and plants undergo during life, which had been discovered by Priestley and others:

> Plants decompose the carbonic acid gas of the atmosphere absorb & convert the carbon, & in the same proportion give forth o[x]ygen[.] carbonic acid gas produced by fermentation [&] combustion & respiration can only be consumed by plants which

exude in the same proportion oxygen. … An exchange is made between carbonic acid gass & oxygene gas; the former the result of the destruction of the principle of life & the latter the fuel by which it is nourished.[14]

These notes show that Percy knew what happened when plants breathed in and out, and the vital part they played for human life. He saw the situation as a mutually beneficial exchange between plants and animals. He notes here, too, that carbon dioxide is released into the atmosphere by dead bodies and rotting vegetables, to be used by living plants in photosynthesis. Without the conversion of carbon into oxygen, humans would not be able to breathe. Percy's point here is that death provides for life. There are a number of cycles at work in the natural world, between animals and plants, and between the living and the dead. Life depends upon these cycles. Although these notes are written by Percy Shelley, there are numerous examples in the journals and manuscripts that show the Shelleys reading, learning and writing together. What Percy knew, Mary may well also have known.

There is more direct evidence of Mary's knowledge in *Frankenstein* when the Creature notices the importance of the seasons. On his way to Geneva, as summer returns, and 'the sun had recovered its warmth', he notes that 'the earth again began to look green'. More generally, sunshine is something that cheers and revives him.[15] Imagining the future life he and his companion will share in the 'vast wilds of South America', after Frankenstein has constructed her, the Creature reveals that he considers sunshine to be one of those natural rights that all living creatures should be allowed: 'the sun will shine on us as on man, and will ripen our food'.[16] His enjoyment of natural, bodily sensations, of the warmth and light of sunshine, reminds us that he is a sentient, living being, who has a right to life too.

THE RECOVERY OF PERSONS APPARENTLY DROWNED

In the *Annual Report of the Royal Humane Society* for 1818, the year of *Frankenstein's* first publication, narratives are given of people saved from death by means of the Royal Humane Society's methods. For example, a Mr W. Woodward recounts the attempted suicide by drowning of a woman (identified only as 'S.D.') in London. He was on his way to Limehouse when he saw a man rush in to rescue someone from a pond in front of him. He helped bring out the body of the woman, 'apparently dead', and proceeded to carry her in 'a horizontal position' to the nearest house they could find.[17]

Woodward ordered the woman to be undressed and made dry; she was deathly cold and there was no fire to warm her. Instead he filled two bottles with hot water and placed them on her feet and chest. What followed sounds like a rudimentary kind of CPR, though Woodward pressed on the woman's stomach (rather than chest) while someone else stopped the women's nostrils as he pressed. They opened her nostrils as Woodward released the pressure on her stomach. They did this for half an hour until the woman groaned, an indication that she was alive. Woodward then gave her a spoon of brandy.

The saved woman's first words to him were 'Let me die', but her wish was not granted. She was conveyed to the local poorhouse where, Woodward tells us, she was well known to be an 'abandoned character'.[18] The case is a reminder that – as Mary Wollstonecraft claimed in her suicide attempts – not everyone wanted to be saved from death (*see Introduction*). Frankenstein contemplates suicide on more than one occasion: 'I was overcome by gloom and misery, and often reflected I had better seek death than remain miserably pent up only to be let loose in a world replete with wretchedness.' Despite these emotions, he tells Walton it was his fate to be 'doomed to live'.[19] The *Annual Report* for 1818 records that 216 people were saved by

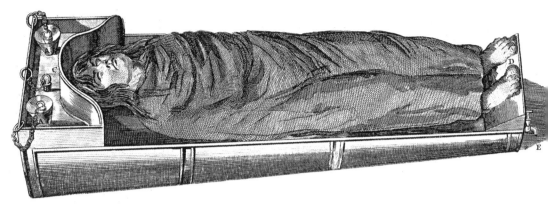

FIG. 6 A body covered with a blanket lying on a heated bath for the purpose of resuscitation, etching, 1790.

Royal Humane Society methods in the past year and that 41 of these had been attempted suicides. Of these, 38 were women, and of the total 41 suicide attempts 38 were 'restored' to life (we do not know the sex of those 'restored'). The Society claimed that none of these people attempted suicide a second time, a fact they attributed to the religious counsel and Bibles they offered to those they saved (see Plate 17).[20]

It might seem surprising to us now that drowning was such a common accidental death in the Romantic period, but at that time much business was done on the canals, rivers and seas. Not many people could swim, and there were no electric lights by which to follow paths safely. Victor Frankenstein acknowledges the importance of rivers when he notes that the creature avoided them because 'it was here that the population of the country chiefly collected'. There are references in the novel to the fishing community in the Orkney Islands. Walton recounts expeditions with whalers off Greenland and in the North Sea (see Plate 14).[21] Of course, the entire novel is framed by Captain Walton's letters sent to his sister while on his journey across the sea to the Arctic. Frankenstein tells Walton his story on Walton's ship while they are stuck in the ice. The likeliest outcome of their predicament is that they and the ship's crew will be drowned, frozen or starved to death.

One of Victor's favourite pastimes is to go out on a boat and spend hours letting the wind carry him. In Geneva he tells us that, while in a boat, he had often been 'tempted to plunge into the silent lake, that the waters might close over me and my calamities for ever'.[22] After destroying the Creature's female companion on the Orkney Islands, he goes out to sea to discard the remains of the body he had begun to create. He falls asleep and his boat drifts off course. When he wakes, and even when he recounts the story much later to Captain Walton, the thought of death at sea terrifies him, whether he would 'be driven into the wide Atlantic, and feel all the tortures of starvation, or be swallowed up in the immeasurable waters that roared and buffeted around me'. He is certain at this point that this is how he will die: 'I looked upon the sea, it was to be my grave.'[23] Percy Shelley would drown in 1822. Despite the fact that he enjoyed sailing a great deal, he never learned to swim well.

In *Frankenstein* the Creature saves a woman from drowning. He describes her state as dead: 'She was senseless; and I endeavoured, by every means in my power, to restore animation.'[24] He succeeds, and his description here is of someone who has been brought back to life. At this time it was believed that deaths by drowning might be easier to return from, that the difference between a living and drowned body was less than between other states. The Royal Humane Society was set up specifically to promote methods thought to be particularly useful in cases of drowning. There was a general understanding that deaths caused by the loss of air (such as drowning, strangulation or suffocation) could be treated in the same manner. Partly because bodies were not disfigured, these deaths were thought to be the easiest to reverse.

When in the novel Henry Clerval's body is found, the villagers assume that he has died by drowning. In fact, he has been strangled by the Creature. William Frankenstein (Victor's younger brother) and Elizabeth (Victor's fiancée) are also murdered in this way. After Clerval's death, Frankenstein dreams that the Creature is strangling him. The Creature clearly chooses to murder his victims using this method. The villagers attempt to revive Clerval using methods publicized by the Royal Humane Society: they 'put [his body] into a bed, and rubbed it; and … went to the town for an apothecary, but life was quite gone'.[25] They are attempting to recover the 'animal heat' thought essential for life. At the request of the Royal Humane Society, a number of medical men published advice on how to revive victims of drowning. The surgeon John Hunter, who was so important in the debate between William Lawrence and John Abernethy (*see Chapter 4*), for example, wrote an essay entitled 'Proposals for the Recovery of People Apparently Drowned'.[26]

The physician James Curry attended Claire Clairmont, in 1814, and Percy Shelley, in 1817, in a medical capacity.[27] In his book *Observations on Apparent Death from Drowning, Hanging, Suffocation by Noxious Vapours…*, Curry established the difference between 'absolute' and 'apparent' death.[28] He promoted accounts such as that of 'Mrs. Page, of Hornsey', who, using Royal Humane Society methods, 'recovered a young girl, who had been taken out of the New River, to all appearance dead'. It was, he writes, 'fully *half an hour* before any signs of life could be observed', but she was revived. Another instance, of a boy who had fallen into a pond in Staffordshire, resulted in success after two and a half hours' 'assiduous employment of the means usually recommended'.[29]

One problem was that the 'signs of life' were not as well known then as they are to us now. Curry thought that in both cases of absolute and apparent death, for example, breathing stopped, the heart ceased beating and the victim was insensible. The only difference, he maintained, was that in cases of absolute death the vital principle is 'completely extinguished', whereas it is merely 'dormant' in cases of apparent death. The difficulty of deciding which kind of death is being dealt with led, he claimed, to recovery after interment. He concluded rather gruesomely that perhaps the only '*unequivocal proof of death*' is the '*beginning putrefaction* of the body'.[30] This demonstrates how uncertain even medical practitioners were in concluding that a person had died.

Curry knew that oxygen was essential to the blood flowing through the body and to the working operation of the heart. He called it 'pure air' or 'vital air' in *Observations on Apparent Death*. In this book he claims that the study of the nature of air is the most important branch of science to medicine.[31] When he offers advice on how to recover the apparently drowned, he describes a process he calls artificial breathing, where the lungs are inflated with fresh air. Figure 7 shows how a long, flexible tube and silver canula should be used with 'other contrivances for preserving persons from danger of various kinds'.[32] Curry encouraged people to continue efforts such as this for six hours before giving up and accepting defeat.

One of the founding members of the Royal Humane Society, the English physician William Hawes, was a firm believer in the efficacy of tobacco enemas, otherwise known as glysters, in cases of apparently drowned persons (*see Plate 20*). Rectal infusions of tobacco smoke were thought to provide warmth and stimulation, deemed the most necessary qualities for life. This method was considered so successful that it became a standard means by which to attempt the revival of

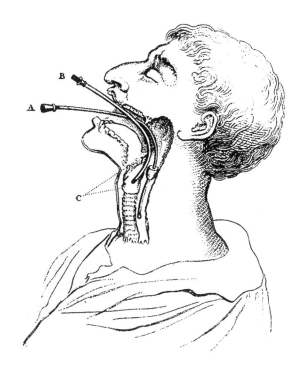

FIG. 7 From James Curry, *Observations on Apparent Death from Drowning, Hanging, Suffocation by Noxious Vapours, Fainting-Fits, Intoxication, Lightning, Exposure to Cold, &c., &c. and an account of the proper means to be employed for recovery*, 1815.

persons who had drowned. The Royal Humane Society routinely issued cards advising people what to do if they found someone in need of help, and the application of tobacco smoke to the intestines was promoted as one of the most efficacious means of aid. We do not know which means were used to revive Mary Wollstonecraft after her attempt to commit suicide by drowning, but such methods as the tobacco glyster might be referred to as being 'inhumanly brought back to life and misery'.[33]

HANGING AND STRANGLING

We know now that there is a chemical process involved in asphyxiation, which is caused by the body being unable to oxygenate tissue.[34] But in the Romantic period hanging, strangling and suffocation were considered purely as processes that killed due to the loss of oxygen.

VITAL AIR   53

For this reason the Royal Humane Society advocated the same treatment to these victims as for drowned persons.

The 1752 Murder Act added the punishment of dissection by surgeons to that of hanging for murderers. Bodies were taken to the Surgeons' Hall in the Old Bailey where they were publicly anatomized. But there were many stories of hanged criminals waking up on the surgeon's table just as the knife was about to cut them open. In one famous case from a century before, Anne Greene of Oxford had been sentenced to hanging for concealing a pregnancy and the subsequent death of her baby. She was a 22-year-old domestic servant, who disclosed she had been 'led ... into fornication' by the grandson of her master.[35] She had miscarried and then attempted to hide the stillborn foetus. As was common during a public hanging, her friends and family held on to her legs in an attempt to shorten the period of her suffering. She was cut down after half an hour and conveyed in a coffin to the University of Oxford's Anatomy School. While her body was being prepared for dissection, it was noticed that she was still breathing and with the efforts of the surgeons she was brought round again.

The methods that the Royal Humane Society would promote were used on Anne Greene. An unnamed woman was employed to lie in bed with her and rub her body, to try to warm her. After her recovery, and with the coffin as a souvenir, Anne received numerous visitors, to whom her father charged admittance. She was pardoned, and eventually married and had three children. She lived until 1659. Other stories did not end so well. Some who survived were hanged a second time the next day; others had their sentence commuted to transportation; others died a few days later. Spectacular recoveries were so well known, though, that friends and family waited outside Surgeons' Hall to see whether their loved ones would revive under the anatomist's scalpel.

There are reports of other so-called 'half-hanged' coming to in their coffins. These much-publicized cases and the reported success of Royal Humane Society techniques contributed to the public's anxieties that they would be buried alive. A number of measures were taken to ensure that people were dead before they were buried, such as waiting a period of time before interment. There are examples of people instructing in their wills that their feet be scratched with a razor or lancet to make absolutely sure that they were dead.[36] Bells were fitted into coffins and burials were supervised.

In Manchester the brother of a wealthy woman called Hannah Beswick was almost buried prematurely. A mourner at his funeral noticed that his eyelids were flickering just as the coffin lid was about to be closed. After investigation he was revived and lived for many years after this date. The experience left Hannah so worried about a premature burial that she wrote into her will that she would always be above ground. When she died, Hannah left her whole estate to her doctor, Charles White, who embalmed her. He checked every day, according to her wishes, that she was definitely dead. Even men of science worried about such things. Humphry Davy specified in his will that his body not be buried for ten days and refused to allow an autopsy. He had, as his brother John described, 'a horror of being buried alive, before animation was completely extinct'.[37]

By the time that Walton meets Frankenstein, there is no notion of the latter's life taking a turn for the better. Frankenstein has accepted this and determined upon death. Walton tells his sister: 'I would reconcile him to life, but he repulses the idea.'[38] Frankenstein mentions feeling suicidal for much of the narrative. It is unclear whether he had decided how he would go about it, but there are times when he needs to be under constant supervision: 'At these moments I often endeavoured to put an end to the existence I loathed; and it required unceasing attendance and vigilance to restrain me from committing some dreadful act of violence.'[39] It may be that Mary intended us to imagine that he attempted to hang himself.

We might assume that, in the novel, Justine Moritz's execution is death by hanging, because this was the punishment for murder in Britain at the time. Frankenstein alludes to this punishment when he considers his fate after being arrested for the murder of Clerval in Ireland: 'Who could be interested in the fate of a murderer, but the hangman who would gain his fee?' The Creature kills all his victims by strangling them. When William Frankenstein dies, we are told, 'the print of the murderer's finger was on his neck'. Clerval's death is accounted for in the same manner: 'He had apparently been strangled; for there was no sign of any violence, except the black mark of fingers on his neck.' The corpse is not disfigured in death other than by the marks caused by the Creature's fingers. When Elizabeth is killed by the Creature similar language is used: 'The murderous mark of the fiend's grasp was on her neck, and the breath had ceased to issue from her lips.'[40] Frankenstein also notes here that Elizabeth is no longer breathing, that crucial characteristic of life. In his book *Observations on Apparent Death*, Curry reports that 'in Hanging, as in Drowning, the *exclusion of Air from the Lungs* is the immediate cause of death.'[41] He proves this with a harrowing experiment involving a dog, which shocks modern, anti-vivisection sensibilities.

As can be seen in Joseph Wright of Derby's painting and in Priestley's and Curry's accounts of their experiments, the use of animals and birds was common. Frankenstein alludes to experiments on living animals when describing the work that led to his creation: 'Who shall conceive the horrors of my secret toil, as I dabbled among the unhallowed damps of the grave, or tortured the living animal to animate the lifeless clay?' He acknowledges the horror of this

FIG. 8 Grotta del cane ('cave of the dog'), Naples: a natural cave which emitted carbon dioxide and was used for experiments on dogs. Engraving c.1700.

work when he tells Walton that 'often did my human nature turn with loathing from my occupation'.⁴² It is telling that in moments of 'insanity' after he has heard the Creature's story, Frankenstein is tormented by visions of the reverse situation. His guilty feelings about vivisection are laid bare in these visions where the animals get their revenge: 'Can you wonder … that I saw continually about me a multitude of filthy animals inflicting on me incessant torture, that often extorted screams and bitter groans?'⁴³

Dogs were often used in experiments to prove that oxygen was the vital component of life. In Naples, Italy, a natural cave that emitted carbon dioxide known as the Grotta del cane (cave of the dog) was used to show the relative properties of gases. Dogs collapsed in the cave because they breathed in the low-lying gas, which humans did not, but then the dogs would revive in the open air. Curry mentions this tourist attraction in his *Observations on Apparent Death*: 'Dogs are usually the subject of this experiment, which is often repeated to gratify the curiosity of travellers.'[44] The cave was visited by the Shelleys in February 1819 but they refused to witness this demonstration, just as the girls try to turn away from the air-pump experiment in Wright's painting. Percy wrote to Thomas Love Peacock: 'The Grotto del Cane too we saw, because other people see it, but wd. not allow the dog to be exhibited in torture for our curiosity.'[45] We can tell something of the Shelleys' views on the use of animals in scientific experiments in this refusal.

The animal rights movement began in this period. The RSPCA started life as the Society for the Prevention of Cruelty to Animals in 1809 and became a Royal Society in 1824. The Creature in *Frankenstein* is made up of both human and animal bodies. He feels an immediate empathy with the birds he hears when he first becomes conscious and tries to imitate their song. He is also a vegetarian, refusing to kill 'the lamb and the kid' merely to 'glut' his appetite.[46] Frankenstein even calls the Creature an 'animal', though of a 'strange nature'.[47] Does the Creature have the same kind of life as humans have? Does he have a soul? He certainly seems to have the needs and desires that humans have. He needs air, water, food and love. The Ancient Greeks thought of the soul as the spiritual breath of the body. In ancient Mesopotamia the soul was thought to reside in the breath and to leave the body with its final breath.

When the Creature contemplates how his existence will change after death, he highlights a number of qualities that are identified as 'vital' in the scientific and medical writings of the day. He emphasizes the loss of his senses, sight, hearing, touch:

> I shall no longer see the sun or stars, or feel the winds play on my cheeks. Light, feeling, and sense, will pass away; and in this condition must I find my happiness. Some years ago, when the images which this world affords first opened upon me, when I felt the cheering warmth of summer, and heard the rustling of the leaves and the chirping of the birds, and these were all to me, I should have wept to die; now it is my only consolation.[48]

Life is typified by light and warmth. The opposites of these, implied by death, are darkness and cold. Death is a consolation to him now when once he would have regretted it.

Why did the Creature choose to strangle his victims? The fact that this method was specified alerts the reader to Mary Shelley's knowledge of the Royal Humane Society's and others' conviction that breath is life. The Creature is aware of the preciousness of air to human life and to his own life. Air also symbolizes liberty and good health. The Creature's revenge upon Frankenstein is to deprive those he loves of this vital element. At the time, it seemed that bringing people back from deaths that were caused by the loss of air was tantalizingly possible. As we shall see, electric-shock treatment, in particular, seemed to have the potential to bring these people back to life.

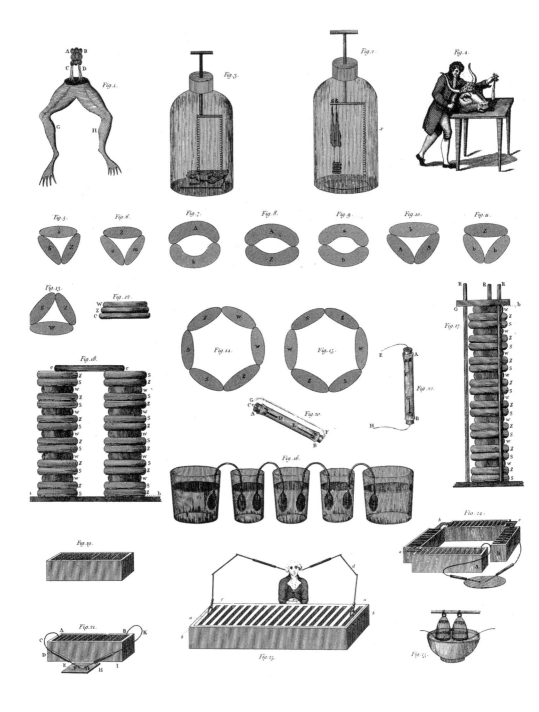

THREE
ELECTRIC LIFE

Victor Frankenstein describes a formative experience when, aged 15, he witnessed a lightning strike 'utterly' destroy an oak tree near his family home. When he asked his father what caused this spectacle, the response was 'Electricity'. Encouraging his son's interest, Alphonse Frankenstein 'constructed a small electrical machine, and exhibited a few experiments; he made also a kite, with a wire, and string, which drew down that fluid from the clouds'. The 'power' of electricity is so impressive that Victor's early interest in ancient alchemy is finally overthrown.[1]

While his first experience of electricity foregrounds its capacity for destruction, there is also some evidence in the novel to suggest that electricity is the means by which the Creature's life is given. Certainly, the cinematic opportunities afforded by lightning and bolts of electricity have been taken up and exploited by numerous film versions of the text, from James Whale's *Frankenstein* (1931) to Kenneth Branagh's *Mary Shelley's Frankenstein* (1994) and beyond. But Mary Shelley does

FIG. 9 Early-nineteenth-century engraving of electrical equipment, batteries and frogs' legs.

not explicitly tell us how the Creature is made. This is, indeed, one of the novel's many strengths. It is one reason why there are so many versions and adaptations of it. The absence of explanation allows us to read our own contemporary anxieties about science and technology through the novel. Mary invents a plausible narrative reason why Frankenstein cannot tell even Captain Walton exactly how he managed to create the Creature. This 'secret', he explains, must not be told because it is too dangerous to pass on to others.[2]

It is not until the 1831 Introduction to the novel that Mary Shelley explicitly suggests that electricity might be the key to Victor's act of creation. She writes: 'Perhaps a corpse would be reanimated; galvanism had given token of such things: perhaps the component parts of a creature might be manufactured, brought together, and endued with vital warmth.'[3] When Mary uses the term 'galvanism', she is referring to the electricity that Luigi Galvani thought he had discovered. This passage shows that Mary knew that the key to life, or what she refers to as 'vital warmth', might be achieved by means of electricity. Despite there being no explicit mention of the use of electricity in the creation scene, there are tantalizing references – which could also be metaphorical – to a vital 'spark' throughout. Electrical shocks had been a commonly used treatment recommended by the Royal Humane Society since 1789 in the recovery of persons drowned. The society's motto reflects this: *lateat scintillula forsan*, 'a small spark may perhaps lie hid'. Life, or the vital principle, had been imagined as a spark since ancient times. The word also reminds us of Prometheus stealing fire from the gods to give to humans.

It is impossible to know Frankenstein's exact method, but at the moment of creation he uses a language that might suggest electricity: 'I collected the instruments of life around me, that I might infuse a spark of being into the lifeless thing that lay at my feet.' Electricity was thought to be a subtle fluid at this time. This is how Frankenstein thinks of it too, so the idea of 'infusing' it into the otherwise dead and inert body makes sense. Later in the novel, Frankenstein uses the same terms when he wishes to 'extinguish the spark which I so negligently bestowed'. At the end of the novel, the Creature poignantly asks why he did not kill himself when the De Lacey family rejected him, repeating the language that his maker had used earlier: 'Why, in that instant, did I not extinguish the spark of existence which you had so wantonly bestowed?'[4] The fact that a spark can be extinguished easily suggests that the vital principle is like a fleck of fire. It also emphasizes life's vulnerability.

ELECTRICITY IN SCIENTIFIC EXPERIMENTS

The association of lightning with electricity that Alphonse Frankenstein revealed to his son had, by 1816, not long been confirmed. Benjamin Franklin suspected that this was the case and had thought of an experiment that would prove his theory correct, but in Philadelphia, where he lived, there were no buildings tall enough for him to try it out. He wrote to Peter Collinson, a fellow of the Royal Society in London, and this letter was published in the society's *Philosophical Transactions* for 1752.[5] The idea had been suggested to him by reading about lightning striking the top of mastheads on ships. Others tried what he suggested, calling it the 'Philadelphia experiment'. Undeterred by the lack of tall buildings, Franklin had come up with the idea of using a kite instead.[6] He built a kite consisting of two pieces of wood crossed and a silk handkerchief. A foot-long piece of wire was used to conduct the electricity from a charged cloud (*see Plate 19*). Immanuel Kant called Franklin 'the Prometheus of modern times' in an essay of 1755, drawing upon the myth that Mary Shelley would later reference in the subtitle to her novel, *Frankenstein; or, The Modern Prometheus*.

She may well have been thinking about Franklin when she imagined Frankenstein.

Percy Shelley's university friend Thomas Jefferson Hogg records that Percy had in his rooms at Oxford 'an air-pump, the galvanic trough, a solar microscope, and large glass jars and receivers'.[7] The 'galvanic trough' would have enabled him to conduct electrical experiments. Percy's 1820 notes from Humphry Davy's *Elements of Agricultural Chemistry* (1813) reveal that he understood electrical charge and the role that electricity could have on encouraging plant growth: 'Electy probably has great effect on plants. Corn sprouted more rapidly in water positively electrified by the Voltaic battery <than in> negatively. The clouds are usually negative – the Earth therefore positive —'.[8] He may have learned about electricity from Adam Walker, an itinerant lecturer who taught Percy at Syon House school. Walker thought that all forms of electricity, light and heat were modifications of the same single principle, which he preferred to call 'fire'. In Walker's model, fire in its 'elementary state' is electricity, and the sun is the source of electricity.[9] When Percy's school friend Thomas Medwin came to visit him at Oxford in November 1810, he recalled that 'He had not forgotten our Walker's Lectures, and was deep in the mysteries of chemistry.'[10] For some people at this time, Walker's single principle was identical to the 'vital principle' that was essential for life.

Alessandro Volta is credited with inventing the electric battery, called the Voltaic pile, in 1799. The unit of electricity called the 'volt' is named after him. He proved that an electric current could be generated by the connection of two different metals. The ability to produce this current and to maintain it at a steady level paved the way for many other important experiments. One of these, undertaken by William Nicholson and Anthony Carlisle, both of whom William

Godwin knew as friends, involved the construction of the first Voltaic pile in Britain. With this new instrument they decomposed water into its two constituent parts, oxygen and hydrogen. Electrochemistry was thus born. Davy, in turn, used these means to isolate nine newly discovered elements, including potassium, sodium and magnesium.

After Frankenstein experiences the revelation in his scientific thought caused by witnessing the tree destroyed by lightning and his father's subsequent electrical experiments, he attends 'a course of lectures upon natural philosophy'.[11] In notebook A, a surviving draft of the novel, there is a clear link between seeing the tree destroyed by lightning and Frankenstein's decision to study chemistry. In this draft, Frankenstein tells Walton: 'The catastrophe of the tree excited my extreme astonishment and caused <induced> me to aply with fresh diligence to the study of chemistry *<natural philosophy>*.[12] This is one of many occasions in the manuscript where Mary Shelley's 'chemistry' is replaced by Percy's 'natural philosophy'. The latter term encompasses more than simply chemistry and was the one that Davy, for example, preferred to use to describe his work. For Davy, the category of 'natural philosopher' referred to someone with a wider perspective than others, a man sensitively alive to the metaphorical, literary and philosophical ramifications of his scientific discoveries. Despite this, Mary clearly envisaged Victor, initially at least, as a chemist specifically.

Mary tries, through a number of scenarios, to account for Frankenstein's love of chemistry. At one point in the manuscript, he says: 'I used when very young to attend lectures of chemistry given in Geneva and although I did not understand them the experiments never failed to attract my attention.' In another attempt, Frankenstein attends the lecture of someone he meets at Clerval's father's house, a certain Monsieur 'O. P.', whom he describes as 'a proficient in Chemistry'.

Intriguingly, Percy experiments with the idea that Frankenstein's 'utmost wonder' is engaged not by the demonstration of an air pump (*see Chapter 2*) but an '*electrical machine*', though this idea was not taken up by Mary.[13]

In the version of the novel that was published in 1818, Frankenstein attends only the final part of a course of lectures 'upon natural philosophy' and is unable to understand its content because of all that he has missed. As a consequence, he is repelled by the discipline: 'The professor discoursed with the greatest fluency of potassium and boron, of sulphates and oxyds, terms to which I could affix no idea.'[14] In the draft notebook, two other elements are mentioned and crossed out: 'zinc bismuth'.[15] Both of these metals had been known since ancient times, and both were important in the practice of alchemy. The electrochemical properties of zinc were also important for both Galvani and Volta. In the first published edition of the novel, however, these elements are rejected.

The two elements that were retained, potassium and boron, were both isolated by Davy, using the new science of electrochemistry, in 1806 and 1807 respectively. The language used by the lecturer, which Frankenstein does not understand, results in his feeling 'disgusted with the science of natural philosophy' though he still reads natural history.[16] In the draft notebook, the lecture seemed to him 'to contain only words' and 'From this time untill I went to Colledge I entirely neglected my formerly adored study of chemistry <*the science of natural philosophy*>.' Again, Mary's choice of 'chemistry' is overwritten by Percy's 'natural philosophy': 'I had a contempt for the uses of modern chemistry <*natural philosophy*>.'[17]

It can be argued, then, that Mary Shelley was thinking about electricity as the means by which the Creature would be brought to life. Even though Victor Frankenstein is not a medical doctor, but primarily a chemist, Davy's career demonstrates that electricity was the chemist's tool (*see Introduction*). Indeed, the use of electricity made Davy famous. Frankenstein mentions 'chemical instruments' or 'chemical apparatus' five times in the novel. Mary Shelley may well have imagined that these would include an electrical machine like the one that Percy himself owned. His sister, Hellen Shelley, recollected that, 'When my brother commenced his studies in chemistry, and practised electricity upon us, I confess my pleasure in it was entirely negatived by terror at its effects.'[18] His friend Hogg tells us that Percy's books were like those read by Frankenstein, both concerned with 'magic and witchcraft' and 'those more modern ones detailing the miracles of electricity and galvanism'. At Eton, Hogg records, Percy 'possessed an electrical machine' and 'contrived a galvanic battery'. Hogg mocks the young Percy Shelley at Oxford when he imagines what wonders might be achieved through the use of electricity. Envisioning a huge galvanic battery, he asks: 'what will not an extraordinary combination of troughs, of colossal magnitude, a well-arranged system of hundreds of metallic plates, effect?'[19] In fact, Davy built exactly this kind of battery in 1813, consisting of 2,000 double plates.

ATMOSPHERIC ELECTRICITY

It is perhaps significant to the question of whether the Creature is brought to life using electricity that lightning plays an important role in the novel. Mary Shelley memorably describes a thunderstorm that takes place as Frankenstein has almost reached home on his return from Ingolstadt. He stops for the evening at Sécheron, described in the novel as a village just outside Geneva, which today is primarily a train station for local and regional trains to get to the north of the city. He has to cross the lake in a boat to arrive at Plainpalais.[20] The lightning is so spectacular that when he gets there Frankenstein ascends a low hill to watch it. He walks through the storm and though it sounds terrifying he does not feel terror. The thunder echoes between the mountains: 'the darkness and storm increased every minute, and the thunder burst with a terrific crash over my head.'[21] His vision alternates between moments of pitch-black darkness and flashes of extraordinary lightning. Frankenstein sees this 'noble war in the sky' as a lament and celebration for his brother William and declaims this aloud. It is at precisely this moment that he notices a figure near him:

> A flash of lightning illuminated the object, and discovered its shape plainly to me; its gigantic stature, and the deformity of its aspect, more hideous than belongs to humanity, instantly informed me that it was the wretch, the filthy daemon to whom I had given life.[22]

For the first time, Frankenstein considers the idea that the Creature may have been William's murderer, and instantly he becomes convinced of this. When next he sees the Creature, again illuminated by the lightning's flash, he is improbably 'hanging among the rocks of the nearly perpendicular ascent of Mont Salève'. The episode demonstrates the Creature's superhuman physical abilities. This strongly Gothic scene, enhanced by the weather, increases the terror of the Creature, but it may also suggest an association between the Creature's life and electricity.[23]

Frankenstein knows that clouds are the means by which atmospheric electricity is emitted because his father had demonstrated this, 'drawing' down electricity using Franklin's kite. Percy Shelley's 1820 poem 'The Cloud' makes clear that he also knew about this process. Captain Walton's quest in *Frankenstein* also has a link with electricity; he wants to discover the magnetic North Pole. As he tells his sister, he hopes that his journey will enable him to ascertain 'the secret of the magnet'.[24] While Victor may, ultimately, be coy about the exact means by which he achieved animation, there certainly are a number of references to electricity in the text itself.

ANIMAL ELECTRICITY

One night in 1780 Luigi Galvani discovered that frogs' legs could conduct electricity. He watched as the legs twitched and sparks flew (*Figure 10*). From this moment on, Galvani was convinced that all animals – including humans – possessed a distinctive kind of electricity in their bodies, a force that he named 'animal electricity'. Volta, however, was quick to realize what had really happened: metals had touched during the experiment and produced the electricity. The

FIG. 10 Galvani discovered that frogs' legs could conduct electricity. Illustration from Galvani, *De viribus electricitatis in motu musculari commentarius*, 1791.

animal body was merely a conductor of electricity not the receptacle of another kind of electricity. But Galvini's experiments fired the imagination and seemed to suggest that the vital principle itself could be electricity. His nephew, Giovanni Aldini, took Galvani's experiments further using ox heads and the bodies of hanged murderers in spectacular theatrical events attended by the rich and famous.

The bodies of murderers were ideal for Aldini's experiments because they were usually in the prime of life and their deaths had been sudden. Their bodies were not ravaged by disease or illness. It was necessary, Aldini writes, to procure a 'human body while it still retained, after death, the vital powers in the highest degree of preservation'. In order to achieve this, he was 'obliged … to place myself under the scaffold, near the axe of justice, to receive the yet bleeding bodies of unfortunate criminals'. Aldini describes his experiments on 'two brigands' who had been decapitated in Bologna in 1802. These men, he writes, were young and in a 'robust' physical condition when they died.[25]

Victor Frankenstein speaks of how his 'human nature' often turns 'with loathing from my occupation, whilst, still urged on by an eagerness which perpetually increased, I brought my work near to a conclusion'.[26] Although Frankenstein decides that in order to create the Creature he must become 'acquainted with the science of anatomy', presumably, like Aldini, he was 'little acquainted with anatomical dissection'.[27] Frankenstein does not mention taking any courses at university that would involve dissection. Instead, his studies of physiology and anatomy appear to have been self-directed. He also tells us that anatomy 'was not sufficient'. As he puts it: 'To examine the causes of life, we must first have recourse to death.'[28] Both Aldini and Frankenstein affirm that they had to put feelings of horror to one side in order to continue in their work.

Aldini writes that for these horrid experiments, 'the love of truth, and a desire to throw some light on the system of Galvanism, overcame all my repugnance'.[29] Frankenstein uses nearly the same phrase when he describes the mere thought of creating the female companion. But he is unable to conquer these feelings: 'I was unable to overcome my repugnance to the task which was enjoined me.' Where he had been blinded by a 'kind of enthusiastic frenzy' during the first Creation, he now feels the full horror of the 'filthy process' in which he is engaged.[30] Even during the first Creation, Frankenstein's 'disgust' at what he had done returns as soon as the Creature lives: 'breathless horror and disgust filled my heart'. It is, specifically, the sight of the Creature moving that terrifies Frankenstein. When 'those muscles and joints were rendered capable of motion, it became a thing such as even Dante could not have conceived'.[31] Frankenstein declares that even the sight of a corpse reanimated could not be so terrible: 'A mummy again endued with animation could not be so hideous as that wretch.'[32] The words used evoke the experience of the uncanny, as well as disgust, horror and repugnance. They are words that particularly describe the idea of a corpse moving again after it has been electrified.

Even the Creature is disgusted by the means through which Frankenstein created him. He leaves the laboratory with papers in which Frankenstein 'minutely described … every step [he] took in the progress of [his] work' and thereby learns how he was made. The Creature describes the process as a 'series of disgusting circumstances' that sickens even himself.[33] The *Edinburgh Review* used the same adjective in their review of Aldini: 'Mr. Aldini has often performed his processes on the dead human subject; but the accounts that he gives of his results, are rather disgusting than instructive.'[34] The revivification of a body using an electric shock might be the peculiar aspect of the novel that Mary Shelley describes in her 1831 Introduction as 'so very hideous an idea'.[35]

When the Creature awakens, Frankenstein tells Walton: 'I saw the dull yellow eye of the creature open; it breathed hard, and a convulsive motion agitated its limbs.'[36] This sounds very like the description of Aldini's attempts to resuscitate 26-year-old George Forster, who had been hanged for the murder of his wife and child at Newgate in January 1803. As Aldini records: 'On the first application of the arcs the jaw began to quiver, the adjoining muscles were horribly contorted, and the left eye actually opened.'[37] The 'convulsive' motion of the Creature seems particularly galvanic.

Aldini's experiments were intended to demonstrate the affinity between electricity and the vital principle. When he used the head of an ox that had been recently killed, the effects were spectacular: 'the eyes were seen to open, the ears to shake, the tongue to be agitated, and the nostrils to swell'. He interpreted the ox's actions as being 'in the same manner as those of the living animal, when irritated and desirous of combating another of the same species.'[38] The ox seemed to be angry and ready to fight.

Aldini's experiments on a murderer in Bologna produced terrifying results: 'I observed strong contractions in all the muscles of the face, which were contorted in so irregular a manner that they exhibited the appearance of the more horrid grimaces.'[39] Spectators were horrified by these scenes, which they interpreted as the imminent resurrection of a violent criminal. The *Newgate Calendar* recorded that in London, when Aldini galvanized Forster, 'Mr Pass, the beadle of the Surgeons' Company, who was officially present during this experiment, was so alarmed that he died of fright soon after his return home.'[40] Drawing upon contemporary cases such as these, Mary Shelley, in *Frankenstein*, chose a topic that would terrify her readers.

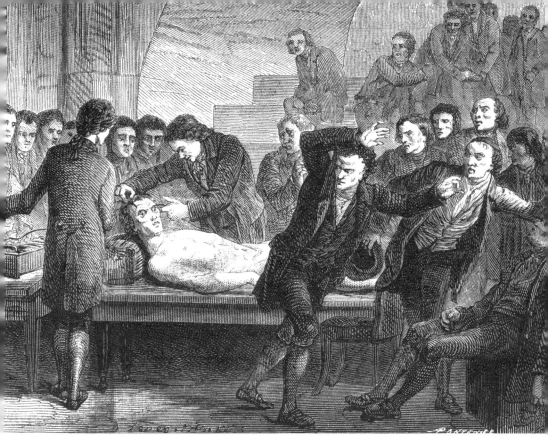

FIG. 11 Dr Andrew Ure attempting to resuscitate the murderer Matthew Clydesdale. Engraving from Louis Figuier, *Les merveilles de la science, ou Description populaire des inventions modernes*, 1867.

The angry manner in which victims 'revived' after electrocution supported the popular belief that a violent criminal was being brought back to life in an unnatural manner. This makes Mary's decision not to use a corpse for the Creature a particularly interesting one. Though he is not made of one single person, the only human bodies to be found legally in the 'dissecting room' are those of hanged murderers. It was specifically in the case of murder that the 'further infamy' of dissection was added to the punishment of hanging (*see Chapter 5*). Together with the animal bodies at the 'slaughter-house', Frankenstein found many of the 'materials' for the Creature's body here.[41]

Other experiments performed by Aldini resulted in corpses seemingly attempting to stand or sit up, open their eyes, clench their fists, raise their arms and beat their hands violently against the table. A number of similar attempts were made to imitate Aldini's successes by British surgeons: in 1803 Dr J.C.S. Carpue attempted, but failed, to resurrect the body of the executed murderer Michael Carney. After the publication of *Frankenstein*, the bid to resurrect the murderer Matthew Clydesdale in 1818 met with greater success: breathing was resumed and similar effects witnessed on the corpse as on that which Aldini had electrocuted. James Jeffray and Andrew Ure, professors at the University of Glasgow, found that 'rage, horror, despair, anguish and ghastly smiles, united their hideous expression in the murderer's face'. Spectators were forced to leave, overcome with 'terror or sickness', and one gentleman fainted.[42]

Aldini was careful to emphasize that his research was motivated by science. His aim was to discover more about the vital powers of the body and he was convinced that electricity was the cause of animation. He urged that further study be made of the influence of galvanism on corpses in cases of accidental and natural deaths. Strongly recommending electrical treatment to the Royal Humane Society as the best way to effect resurrection in cases of drowning and suffocation, he encouraged the use of portable electrical machines at the accident scene.[43] Aldini hoped that 'by pursuing these researches more in detail, they will one day make us better acquainted with the character of the vital powers'.[44] This claim was made for electricity at a time when, as Aldini put it, 'medicine presents ... so few resources'.[45]

For Aldini, animal electricity was synonymous with vitality, the life-giving force that supposedly remained within the body for some time after death. Through electricity, he believed that he might be able to bring back and extend life. His language reminds us strongly of the language of power and control that is used by Frankenstein, for Aldini too is determined: 'to continue, to revive, and, if I may be allowed the expression, to command the vital powers; such are the objects of my researches, and such the advantages which I purpose to derive from the action of Galvanism.'[46] In other words, Aldini intended to use electricity to awaken and resurrect any vitality that still existed in the dead body. Professor Waldman had used this language in the lecture heard by the receptive young Frankenstein at Ingolstadt University. Among the 'new and almost unlimited powers' acquired by the 'modern masters' of science was the ability to 'command the thunders of heaven'.[47] Controlling electricity could mean controlling life.

ELECTRICAL RESUSCITATIONS

John Hunter, a member of the Royal Humane Society board and an important figure in the vitality debate between William Lawrence and John Abernethy (*see Chapter 4*), was an advocate of using electricity where other methods had failed. Others were less sure: Franklin, for example, was a notable sceptic, and thought that any perceived success was only dubious and temporary. Charles Kite in his 1788 *Essay in the Recovery of the Apparently Dead*, which was awarded a silver Humane Society medal, advised the use of electricity in cases of accidental death. He recounts a case in 1785 where a young man had recovered after he had been 'a considerable time under water' and had been 'exposed in his wet cloaths to the cold air for the space of an hour before any means could be used to restore him'.[48]

Every means known to the Humane Society practitioner was used upon the man: 'artificial respiration, warmth, the tobacco glyster, volatiles thrown into the stomach, frictions, and various lesser stimuli were employed near an hour, without the least benefit or alteration.'

Their efforts were without success. At this point, therefore, electricity was applied and shocks were sent through the poor man's body in 'every possible direction'. His muscles evinced 'strong contractions, nearly as violent as is usually observed in healthy people'.[49] These contractions continued for another two hours. It had now been four hours since he had fallen overboard. Though the case did not end happily and the man died, it convinced Kite that electrical shocks were a good test of whether vital powers remained in the body:

> the electrical shock is to be admitted as the test, or discriminating characteristic of any remains of animal life; and so long as that produces contractions, may the person be said to be in a recoverable state; but when that effect has ceased, there can no doubt remain of the party being absolutely and positively dead.[50]

At this time, when death was difficult to verify, electric shocks were proposed as a test to determine absolute death. For Kite and others, electricity seemed to offer a means by which to triumph over 'apparent' death (*see Chapter 2*).

ELECTRICAL THERAPY

Aldini had high hopes for electricity's medical benefits for the living and the 'apparently' dead. He thought that electricity could be used to help hearing and sight problems, in cases of asphyxiation and drowning, and for melancholy and madness.[51] Walker proposed electrical treatment for palsy, cataracts, circulation, removal of obstructions, inflamed eyes, tumours, paralysis, rheumatism, deafness, toothache and numbness.[52] According to Hogg, Percy used electricity to treat his sisters' chilblains, as Walker advised.[53] John Wesley published a compendium of case reports in which electricity had purportedly cured various illnesses.[54] Many of these treatments rested upon the increased circulation and renewed pulse that an electric shock could give the patient. Erasmus Darwin also favoured shock treatment.[55]

A number of medical and scientific figures believed that external atmospheric electricity affected internal animal electricity. Walker, in *A System of Familiar Philosophy*, argued that the weather could affect well-being: 'may not languor, and low spirits, in moist weather, arise from a want of electricity in the air, &c? For easterly winds always induce disorders, and it is notorious that the air betrays less signs of electricity in those winds than in any other.'[56] In Walker's pathology, we need a certain amount of electricity to feel well. Any increase or decrease in the atmosphere's electricity produces a corresponding response in our animal electricity. Changes in the atmosphere affect changes within us.

As is perhaps clear already, electricity in the late eighteenth and early nineteenth centuries was hailed as a cure-all for innumerable illnesses and diseases, and was used to treat a wide range of problems. While some embraced it, others worried that electrical therapy was a sham. Godwin translated the French commissioners' report that detailed the many ways in which animal magnetists duped their gullible patients.[57] In the 'Historical Introduction' written to accompany the *Report*, Godwin is unequivocal: 'it can no longer be concealed that the system of the animal magnetism is to be regarded as an imposture.'[58] Mary Wollstonecraft warned women against the 'fashionable deceptions' of the animal magnetists in *A Vindication of the Rights of Woman* (1792). She was prepared to consider the existence of a 'subtle electric fluid' within our bodies because, she reasoned, 'the most powerful effects in nature are apparently produced by fluids, the magnetic, etc.' If this is the case, she asks, why might not 'the passions … be fine volatile fluids' too?[59] Wollstonecraft's query demonstrates how open and unresolved such questions were. Lack of knowledge led to extraordinary claims for electricity as the vital principle.

Percy Shelley certainly believed in the efficacy of animal magnetism. There was a rumoured family connection between Percy's grandfather, Bysshe, and one of the most well-known figures to use electricity in medical treatments. Bysshe 'was believed to have been a partner in the professional activities of Dr. James Graham, the notorious mesmeric charlatan.'[60] Graham had studied medicine at Edinburgh University, where John Polidori would later also train (*see Chapter 1*). There is some doubt about whether he qualified in Edinburgh, but after his time there he assumed the title of Doctor. He travelled in America, practised as a doctor and was influenced by Franklin's electrical discoveries.

In 1774 Graham returned to England from America and in 1775 began practising medicine in London. Electrical treatments featured heavily in his cures. Despite his reputation as a quack, his practice became very fashionable, particularly after he treated minor aristocracy in Paris in 1779. In autumn of the same year he established himself in London, on the Royal Terrace, Adelphi, facing the Thames. He bought a large house, which he decorated elaborately and filled with expensive equipment, calling it the 'Temple of Health'. The walls of the entrance hall were lined with crutches supposedly thrown away by the patients that he had cured. Here, he asserted the healing properties of the earth and promoted 'earth-baths', which involved being buried naked in the ground. The Temple housed a 'celestial bed' that purportedly cured infertile couples and a 'magnetic throne' that emitted electrical shocks. Before she became Nelson's mistress, Graham's beautiful assistant in the 'Temple of Health' was none other than the young Emma Harte.[61]

The Aldephi house became too expensive to maintain and in the spring of 1781 Graham was forced to move to Pall Mall. He lowered his charges (a further indication that he had fallen from fashion in London society), although a night spent in the celestial bed still cost £50. He began to travel again and was imprisoned once for disobeying legal injunctions prohibiting him from lecturing in public. In December 1783 he claimed to know how to live to be 150 years old. His promises to patients became increasingly wild and outrageous, and in his later years he became a religious enthusiast, denounced by some as a madman.

Graham was mimicked in the satirical play *The Genius of Nonsense*, written by George Colman the Elder and performed at the Haymarket Theatre on 2 September 1780, when Graham was at the height of his popularity. In this drama, Graham's character, the 'Emperor of Quacks', affects to bring about every kind of cure with electricity. It is clear from the play's script that Graham prescribed electrical shocks irrespective of what the particular illness was. In the play, the Emperor of Quacks treats a diseased hand, an eye and a swollen foot with these means, all with the same lack of success. In a lecture to the audience, he describes the treatment as follows:

> Electricity, Air, Musick, and Magnetism are my Ministers. The fire is made to pass thro' any particular part of the patient's Body, giving a Number of little pleasant vibratory shocks – in Consequence of which the Blind recover their Eyes, the Deaf their Ears, the Dumb their Tongues, the Lame their Legs.[62]

The list of ailments that he pretends to cure extends to the use of electrical shocks to revive a patient 'at the Point of Death'.[63] The popularity of the play, which had a successful run, demonstrates Graham's fame, though it denounces his medical ability.

By the end of his career, Graham was largely dismissed as a charlatan, though he was by no means the only practitioner to promote the use of electricity as a treatment in all manner of ailments. Nor was he the only one to make a link between electricity and life. He claimed that electricity introduced to the system would 'supply a vivifying spirit – a *pabulum vitae*, to the injured'. Graham's treatment operated at the level of the essential vital principle: 'these vivifying medicines breathing a pure ætherial flame, refresh, restore, reanimate, acting INSTANTLY and powerfully, as a Divine restorative, reviving and recruiting the principle, and strengthening the very stuff of life.'[64]

If the key to creation in *Frankenstein* is electricity, Graham proves that there were others who thought the same.

The link between electricity and life was also made by more legitimate men of science than perhaps were Graham and Aldini. When Davy wrote to Samuel Taylor Coleridge of his early experiments with galvanism, he told him: 'I have made some important galvanic discoveries which seem to lead to the door of the temple of the mysterious god of Life.'[65] In his early essay on nitrous oxide, Davy writes that this is the reason why chemistry is the most important of the sciences: 'Thus would chemistry, in its connection with the laws of Life, become the most sublime and important of all sciences.'[66] Davy's galvanic discoveries would be used in the debate on the nature of life that took place in the Royal College of Surgeons between 1814 and 1819 where, once again, electricity was considered a good candidate for the vital principle.

FOUR
THE VITAL PRINCIPLE

While the 'spark' of life mentioned in *Frankenstein* might refer specifically to electricity, it might also refer to something less precisely defined. Many scientific and medical writers and practitioners thought there was a 'vital principle', a 'something' added to the material body, which made it move and live. Victor Frankenstein speaks in these terms: 'Whence, I often asked myself, did the principle of life proceed?' He tells Captain Walton: 'After days and nights of incredible labour and fatigue, I succeeded in discovering the cause of generation and life; nay, more, I became myself capable of bestowing animation upon lifeless matter.'[1] Frankenstein's language here reveals that he considers the body of the Creature to be 'lifeless matter' until animation is 'bestowed' upon it.

As we have seen, discussion of what life was and how living bodies differed from dead ones raged in the first decades of the nineteenth century. One sign of this was the very public and direct disagreement between two surgeons at the Royal College of Surgeons, John Abernethy and William Lawrence, which polarized the medical profession and sent ripples through British society. The debate was split along political, religious, nationalist and generational lines. Abernethy was older and more conservative; Lawrence the young, radical firebrand, railing against his superiors. Their views entered the public domain with the printing of their lectures, which were reviewed in the journals of the day, and further books and pamphlets were written criticizing the positions taken by the key players.

Lawrence, eventually, lost his hospital post, was forced to recant his ideas, and was refused copyright of the publication of his 1819 lectures. He was accused of 'denying Christianity and Revelation, which was contrary to public policy and morality'.[2] Lawrence's writings were likened to Percy Shelley's *Queen Mab* and Byron's *Cain* by their detractors. Frankenstein's language, and the way that he creates and animates the Creature, makes it clear that he considers life to be as Abernethy argued. Frankenstein was thus on the side of the old-fashioned, dogmatic and orthodox figure in the debate. The Shelleys were personally acquainted with Lawrence (very much the underdog in the controversy), so it is likely that, by associating Frankenstein with Abernethy, Mary was in fact criticizing both.

SPONTANEOUS GENERATION

The nature of life had been the subject of scientific investigation before Abernethy and Lawrence took to the stage. Mary recalled an experiment that had been attributed to Erasmus Darwin, 'who preserved a piece of vermicelli in a glass case, till by some extraordinary means it began to move with voluntary motion', in her 1831 introduction to the novel.[3] The literal meaning of 'vermicelli' is 'tiny worms'.[4] While Darwin may not have performed this actual experiment, he had noted how 'Microscopic animals [were] produced from all vegetable and animal infusions' in his note on 'Spontaneous Vitality of Microscopic Animals' in *Temple of Nature*.[5] There had been much speculation in decades previous to this on where the animalcules came from that seemed to generate spontaneously from rotting meat. Mary seems to reject the idea that these creatures offer an idea of how life originates when she continues her introduction with 'Not thus, after all, would life be given.'[6] Others continued to question what life was, though, if it could be witnessed in such creatures. What did human life have in common with the life of mites and flies?

The development of the microscope had led to sightings of tiny creatures that could not be seen with the physical eye. In the 1740s John Tuberville Needham was convinced that he had seen microscopic creatures which were capable of self-generation. For a number of men of science, the ability to reproduce was the vital principle. For example, in Georges-Louis Leclerc, Comte de Buffon's 1780 *Natural History*, a clear distinction is made between organic and inorganic matter; the latter is 'perfectly inert, and deprived of every vital or active principle'.[7] Buffon thought the ability to reproduce was peculiar to animals and vegetables. Lazzaro Spallanzani refuted this idea, demonstrating that boiling meat, or ensuring the container that meat was in was hermetically sealed, would mean that microbes (as they would now be called) did not appear.

There were a number of competing ideas about what the vital principle was at this time. Johann Friedrich Blumenbach thought it was a formative drive, *Bildungstrieb*, the impulse of organic matter to form itself. He believed that life had a peculiar power, which is distinct from the *'common properties of dead matter'*.[8] This vital principle is the *'chief principle of generation, growth, nutrition and reproduction'*.[9] Samuel Taylor Coleridge was inspired by Blumenbach's idea to explore it as a metaphor for the imagination: 'I wish much to investigate the connection of the Imagination with the Bildungstrieb.'[10]

There was a general sense that all living creatures held a place in the great Chain of Being, a hierarchical order that progressed from the mineral world, through vegetables, to human life and finally reached the divine. Humankind was placed firmly at the top of all mortal, living beings. The complexity of life was supposed to be represented by this chain of being, which moved from simple animalcules merely able to evince irritability, to the most sensible, or sensitive, creatures. Much attention was paid to the moments of gradation on the Chain of Being. For example, mimosa (*Mimosa pudica*), also called the sensitive plant, was placed at the juncture between plants and animals (*see Plate 23*). This plant retracts its leaves when touched and seems, in doing so, to have a kind of conscious life. Percy Shelley wrote a poem titled 'The Sensitive Plant'. He, with other of the Romantic poets, had a more egalitarian view of life than was demonstrated by the Chain of Being. They thought that life was shared by animals and plants, rather than that there was a kind of life exclusive to humans. Certainly the question of how to categorize something as living was one that continued to perplex. On 22 October 1821, Percy wrote from Italy: 'There are also curious fleshy flowers, & one that has blood & that the peasants say is alive.'[11]

Thomas Medwin remembered Percy using his solar microscope to examine 'The mites in cheese, where the whole active population was in motion – the wings of a fly – the vermicular *animalculae* in vinegar, and other minute creations still smaller, and even invisible to the naked eye.'[12] In his essay 'On the Devil and Devils', Percy provokes his reader with: 'the Devils may be like the animalculæ in mutton broth, whom you may boil, as much as you please, but they will always continue alive and vigorous.'[13] As this demonstrates, there was much debate concerning what life was, which beings could be said to be living, and whether humans shared life with forms as simple as these microscopic creatures, insects, plants and animals. The different ideas of life expressed were motivated by political, religious and national sentiments. Mary's novel emerged from late-night discussions on 'the nature of the principle of life, and whether there was any probability of its ever being discovered and communicated'.[14] John Abernethy thought that the late surgeon John Hunter had discovered it and that it was his duty to communicate it.

JOHN ABERNETHY AND WILLIAM LAWRENCE

When he stepped up on the stage to deliver his first anatomical lectures to the Royal College of Surgeons in London in 1814, John Abernethy, Professor of Anatomy and Surgery to the College, first thanked the man who had been his 'instructor', Sir William Blizzard. This was right and proper. He noted particularly the 'excellent advice' Blizzard had given him while in this role.[15] Blizzard was described as the '*beau ideal* of the medical character', who cautioned his students 'never to tarnish its lustre by any disingenuous conduct, by any thing that wore even the semblance of dishonour'.[16] The words could seem as much a warning as friendly advice. Abernethy's own student, William Lawrence, was not to take heed. Lawrence took on and lost the battle against his mentor, who accused him of bringing the whole of the medical profession into disrepute.

By 1814, Abernethy was 50 years old and an establishment figure, although his journey to this point had not been easy. The profession was still very nepotistic and Abernethy claimed that he had been promised an

FIG. 12 Statue of John Hunter, an early advocate of using electricity for resuscitation, by Henry Hope-Pinker, 1886.

FIG. 13 John Abernethy, mezzotint by E. McInnes, after Sir T. Lawrence, 1842.

appointment which never materialized. Surgeons continued in their posts beyond the age of 60 and, in the end, Abernethy did not achieve the status of full surgeon at St Bartholomew's Hospital until 1815.[17] He was a complex character, who does not seem to have cared to make himself very likeable.

He believed that the lifestyles of the rich were responsible for their illnesses and recommended abstemious eating and drinking habits as the answer to all diseases. This idea particularly appealed to Percy Shelley, who wrote 'see Abernethy' in a note to his 1814–15 'Essay on the Vegetable System of Diet'.[18] Abernethy warned about the danger of eating too much meat in a book on the *Constitutional Origin and Treatment of Local Diseases* to which Percy referred.[19] Charles Grove claimed that Percy attended 'a course' of Abernethy's lectures in 1811, after being expelled from the University of Oxford, when he had decided to become a surgeon himself.[20] As a doctor, Abernethy was best known for his refusal to physically examine patients. He became known as 'My Book Abernethy' because he preferred instead to send them to read his *Surgical Observation on the Constitutional Origin and Treatment of Local Diseases*. Though his daughter did not accompany him, William Godwin attended one of Abernethy's lectures in 1813.[21]

Abernethy was opposed to vivisection and avoided operating on patients wherever possible. He was sometimes 'seen in tears after carrying out a major operation', demonstrating just how traumatic surgery was without anaesthesia.[22] Despite this sensitive side, he was also remembered for his rudeness to patients and indifference to their class or rank. After his death, Abernethy was described as plain-speaking, conservative in his politics and religion. He frequently appealed to the 'moral bearings of any subject under discussion', as we see plainly in the Royal College of Surgeons' debate.[23] Abernethy was one of Coleridge's doctors and Coleridge claimed that the former's

Physiological Lectures were 'dictated solely by the writer's [Coleridge's] wishes'.[24]

In contrast, William Lawrence's early career was filled with honours and publications. He was apprenticed to Abernethy aged 16, and, as was common in those days, he lived in Abernethy's house for five years under his tutelage.[25] Only three years after Lawrence's apprenticeship began – and three years earlier than was usual – Abernethy promoted his student to the position of demonstrator for his anatomy lectures. This shows both Lawrence's skill and, presumably, Abernethy's belief in him. Lawrence held this post for twelve years, until 1812.

In 1806, Lawrence won the Royal College of Surgeons' Jacksonian Prize for his work on ruptures. In 1807 he published a translation of Blumenbach's *Short System of Comparative Anatomy*. In 1813 he was elected assistant surgeon at St Bartholomew's Hospital and became a fellow of the Royal Society. He was made surgeon at the London Infirmary for Diseases of the Eye (Moorfield's Eye Hospital today) and was given the sinecure office of surgeon to the Royal Hospitals of Bridewell and Bethlem.

Lawrence became a professor at the Royal College of Surgeons in 1815, a position that Abernethy held simultaneously. This meant that for the next five years they gave alternate introductory lectures to medical students. Both chose the nature of life for the subject of their lectures and they contradicted and opposed each other. Lawrence was the polar opposite of his mentor Abernethy. Rather than being gruff and unwelcoming, Lawrence was polite, charming and erudite. He also had rather alarming links to political radicals such as the satirist William Hone.[26] From at least 1814, Lawrence became the Shelleys' doctor. He was the doctor who told Percy to stop writing poetry and to travel to Italy for the good of his health in 1817. Mary referred to

FRANKENSTEIN;

OR,

THE MODERN PROMETHEUS.

IN THREE VOLUMES.

Did I request thee, Maker, from my clay
To mould me man? Did I solicit thee
From darkness to promote me?——
                    PARADISE LOST.

VOL. I.

London:
PRINTED FOR
LACKINGTON, HUGHES, HARDING, MAVOR, & JONES,
FINSBURY SQUARE.

1818.

PLATE 1 (*previous page*) Title page of the first edition of *Frankenstein*, 1818.

PLATE 2 Portrait miniature of Mary Wollstonecraft Shelley by Reginald Easton, posthumous.

PLATE 3 Portrait of Percy Bysshe Shelley (1792–1822), after Amelia Curran.

PLATE 4 Sir Humphry Davy portrait after Sir Thomas Lawrence, oil on canvas, based on a work of c. 1821.

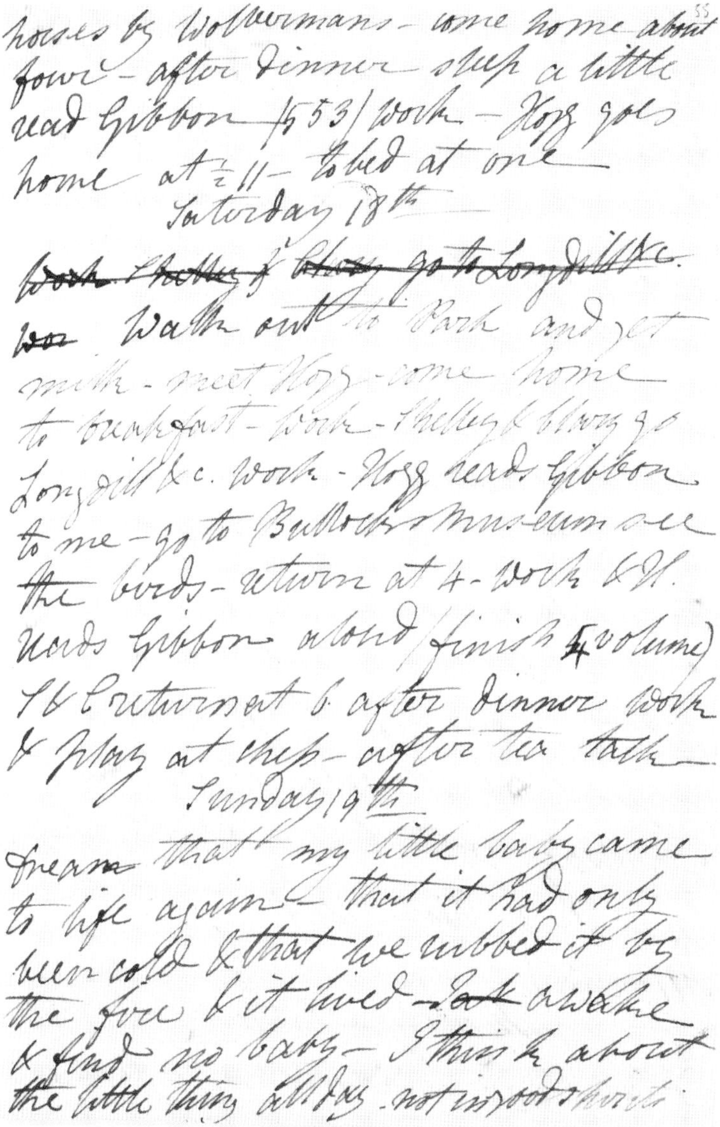

houses by Wolvermans – come home about four – after dinner sleep a little read Gibbon (p 53) work – Hogg goes home at ½ 11 – to bed at one

Saturday 18th

~~Work Shelley & Hogg go to Longdill &c.~~

~~Wor~~ Walk out to Park and get milk – meet Hogg – come home to breakfast – work – Shelley & Hogg go to Longdill &c. work – Hogg reads Gibbon to me – go to Bullocks Museum see the birds – return at 4 – work & H. reads Gibbon aloud (finish 4 volume) S & C return at 6 after dinner work & play at chess – after tea talk

Sunday 19th

Dream ~~that~~ my little baby came to life again – that it had only been cold & ~~that~~ we rubbed it by the fire & it lived – ~~I~~ awake & find no baby – I think about the little thing all day – not in good spirits

PLATE 5 A page from Mary Shelley's journal. She writes 'Dream that my little baby came to life again' in the entry for 19 March 1815.

PLATE 6 (*following spread, left*) A page from the *Frankenstein* manuscripts with Professor Waldman's speech, beginning 'The ancient teachers'.

PLATE 7 (*following spread, right*) A page from the *Frankenstein* manuscripts with the beginning of Professor Waldman's speech, ending with 'with its own shadows'.

but there, at the back of his head, were nearly black. He was short in person but remarkably erect [and] his voice the sweetest I had ever heard. He began his lecture with *by a recapitulation of the* history of chemistry and *made by various* of the various improvements *of learning* men *made* pronouncing *with honour* the names of the greatest discoverers. He then took a cursory view of the present state of *the science*, and explained many of its terms. *after making* [a] few preparatory experiments *he* concluded with a panegyric upon *modern* chemistry the words of which I shall never forget.

"The ancient teachers of this science," said he, "promised impossibilities and performed nothing. The modern masters promise very little. They know that metals cannot be transmuted and that the elixir *of life* is a chimaera. But these philosophers whose hands appear only made to dabble in dirt and their eyes to pore over the microscope or crucible, have indeed performed miracles. They penetrate into the recesses of nature and show how she works

"in her hiding places. They ascend into
"the heavens; they have discovered how
"the blood circulates, and the nature
"of the air we breathe. They have
"acquired new and almost unlimi
"ted powers, – They can command the
"thunders of heaven, mimick the
"earthquake, and even mock the
"invisible world with its own
"shadows".

I departed highly pleased with
the professor and his lecture &
paid him a visit the same evening.
His manners in private were even
more mild & attractive than in public. For
there was a certain dignity in his
manner during his lectures which
was replaced by the greatest affability
affability and kindness in his own
house. He heard my little narration
concerning my studies with attention
smiled at the names of Cornelius
Agrippa and Paracelsus but without
the contempt that Mr Krempe had
exhibited. ~~ended by saying~~ that
his lecture had removed my prejudice
against modern chemists and request-
ed at the same time his advice
concerning the books I ought to procure

He said that these were men to whose indefatigable zeal modern philosophers were indebted for most of the foundations of their knowledge ~~had~~ They left to us, as an easier task to give new names, & arrange in corrected classi fications the facts which they to a great degree had been the instruments of bringing to light The labours of men of genius however erroneously directed scarce ly ever failed.

PLATE 8 Thomas Cole, *Scene from Manfred*, 1833, oil on canvas.
PLATE 9 John Ruskin, *Cascade de la Folie, Chamonix*, 1849, watercolour.

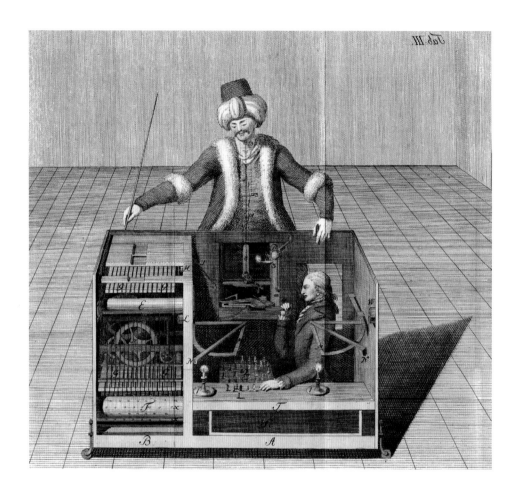

PLATE 10 M. de Kempelen's *Chess-Player* from *Ueber den Schachspieler des Herrn von Kempelen und dessen Nachbildung*, Leipzig, 1789.

PLATE 11 Joseph Wright of Derby, *The Alchymist, in Search of the Philosopher's Stone, discovers Phosphorus, and prays for the successful conclusion of his operation, as was the custom of the ancient chymical astrologers*, oil on canvas, exhibited 1771, reworked and dated 1795.

PLATE 12 Joseph Wright of Derby, *An Experiment on a Bird in the Air Pump*, 1768, oil on canvas.

PLATE 13 (*overleaf*) Percy Bysshe Shelley's manuscript notes on Humphry Davy's 1813 book *Elements of Agricultural Chemistry*.

Sugar & farina mutually convertible into each other.

Sap produced from water, & produces the varieties of substances in plants.

Soil, consists of earths — earths consist of are composed of highly inflammable metals & oxygen; are undecomposed by mere vegetation — (or may be so?)

Water composed of hydrogen & oxygen. Discovd by Cavendish in 1785.

The earth is the laboratory in which the nutriment of vegetables is prepared. Manure is useful & may be converted into organized bodies. Plants decompose the carbonic acid gas of the atmosphere absorb & convert the carbon, & in the same proportion give forth oxygen. Carbonic acid gass produced by fermentation & combustion & respiration can only be consumed by plants which

...in the same proportion oxygen. ...ciates from the progress of the ...of animal life a principle necessary to the existence of vegetables is produced, & from the functions of vegetable existence, animals derive a supply of other substances indispensible to their ... An exchange is made between carbonic acid gas & oxygen gas; the former the result of the destruction of the principle of life, & the latter the fuel by which it is nourished.

Plants consist chiefly of carbon & ...ygen. Manures contain these ...incipals. Their combination forms the result.

Gypsum, alkalies & salts operate not as stimulants to excite absorption, but as producing in themselves an aggregate of substance...

PLATE 14 A whale being speared with harpoons by fishermen in the Arctic Ocean, A.M. Fournier, engraving after E. Traviès.

PLATE 15 Solar microscope, part of an Adams Universal Kit, late eighteenth century.

PLATE 16 Resuscitator kit, c.1800–1850.

PLATE 17 A man recuperating in bed at a receiving house of the Royal Humane Society, after resuscitation by W. Hawes and J.C. Lettsom from near drowning, watercolour by R. Smirke (1752–1845).

PLATE 18 Read-type Medical Frictional Generator, English, c. 1770.

PLATE 19 Benjamin West, *Benjamin Franklin Drawing Electricity from the Sky*, 1816, oil on slate.

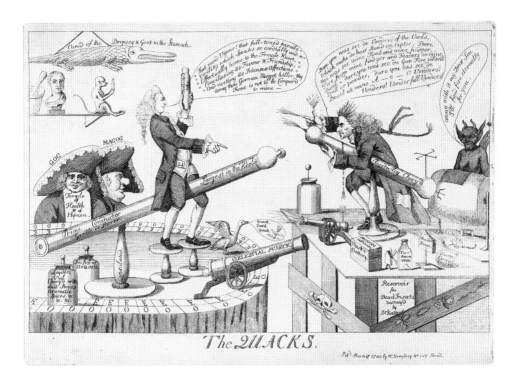

PLATE 20 John Bull being bled and given an enema (clyster) by politicians. *Doctor Sangrado curing John Bull of Repletion – with the kind Offices of young Clysterpipe & little Boney*, by James Gillray, hand-coloured etching, published 1803.

PLATE 21 Contemporary political cartoon satirizing medical treatments by George Cruikshank, *Bleeding & Warm Water! or, The Allied Doctors bringing Boney to his Sense's*, 1813.

PLATE 22 James Graham and Gustavus Katterfelto in combat using electrotherapy machines as weapons. Etching, 1783.

PLATE 23 Illustration of a mimosa (*Mimosa pudica*), a plant that retracts its leaves when touched, from Sydenham Teast Edwards, *The Botanical register: consisting of coloured figures of exotic plants cultivated in British gardens with their history and mode of treatment*, 1825.

PLATE 24 William Godwin's diary; the entry for 4 May 1818 reveals that he saw William Lawrence.

1818

May 3. Su. Life, 3 pp. Boswel, p. 182. Weale & Collins & Baxter dine. ~~Of Religion~~
Of the Old English Writers.

4. M. Life, 1 page. Will. Exhibition; adv. Este, Hill, Hayward, Mulready, Harlow, Dawe, (Gifford) & Perry. Boswel, p. 214. Call on Lawrence, Surgeon.

5. Tu. Life, revise. Write to Jony. Boswel, p. 252. Rodd's, Bohnvright calls: call on Sherwood.

6. W. Life, revise. Boswel, p. 306. Theatre, Douglas.

7. Th. Life, 4 pages. Boswel, p. 338. British Gallery; Cartoons, &c. Baron at tea. Of Religion.

8. F. Life of LK North, p. 168. Boswel, p. 380.

9. Sa. Life, 3 pp. Life of North, p. 210. Sup at Wolcots. William at home, 3 nights.

I cannot get on at all without it I wish you would write to Swinival for more – I have not been out yet this day was too windy & rainy & and indeed the season advances very fast which renders ~~something about~~ A.'s affairs pressing we must decide to do ourselves or send her with in a ~~the~~ month.

It is well that your poem was finished before this edict was issued against the imagination. but my little eclogue will suffer from it – By the bye tas of authorship do get a sketch of Godwins' plan from him – I do not think that I ought to get out of the habit of writing and I think that the thing he talked of would just suit me. & I am glad to hear that J. is well I told you that after what had passed he would be particularly gracious. As to Mrs G. something very analagous to disgust arises whenever I mention her that last accusation of Godwins adds bitterness to every feeling I ~~so~~ ever felt against her.

Send William a present of fruit and a little money. – Pray also dearest do get the state of your accounts from your banker – and also ffor I might as well pack all my commissions into one paragraph send my broach do as soon as you can and as your hair is to be in it have

lock ready cut when you go to the jewellers
& your hair cut in London – For any other
commission be sure to consult your tablets.

Your babes are quite well but I have
had some pain in perceiving or imagining that
Willy has almost forgotten me – and seems to
like Elise better – but this may be fancy &
will certainly disappear when I can get out
and about again – Clara is well and gets
very pretty. How happy I shall be when my own
dear love comes again to kiss me and my
babes – As it seems that your health principally
depends upon care pray dearest take every possi
ble precaution. I have often observed that rain
has a very bad effect upon – if therefore you have
rain in London do not go out in it.

Clare told me to send horse & bag to Maidenhead
to wait for her which I did but she has not come
but I suppose I shall receive an explanation
by tomorrow's post

Adieu – dearest – Come back as soon as
you may and in the mean time write me
long long letters – Your own Mary

Chapter 7th

It was on a dreary night of November that I beheld ~~the from on the~~ my man compleated, ~~and~~ with an anxiety that almost amounted to agony. I collected instruments of life around me ~~and endeavour~~ that I might infuse a spark of being into the lifeless thing that lay at my feet. It was already one in the morning, the rain pattered dismally against the window panes & my candle was nearly burnt out, when by the glimmer of the half extinguished light I saw the dull yellow eye of the creature open — It breathed hard, and a convulsive motion agitated its limbs.

~~But how~~ How can I describe my emotion at this catastrophe, or how delineate the wretch whom with such infinite pains and care I had endeavoured to form. His limbs were in proportion and I had selected his features as beautiful. ~~Handsome~~ ~~Handsome~~ ~~Beautiful~~! Great God! His yellow ~~dun~~ skin scarcely covered the work of muscles and arteries beneath; his hair of a lustrous black, & was flowing and his teeth of a pearly whiteness but ~~these~~ luxuriances only ~~formed~~ formed a more horrid contrast with his watry eyes that seemed almost of the same colour as the dun white sockets in which they were set,

PLATE 25 (*previous spread*) Mary Shelley tells Percy in her letter of 26 September 1817 that she is glad he finished his poem before William Lawrence advised him not to write poetry for health reasons.

PLATE 26 Mary Shelley's famous line 'It was on a dreary night of November' in the *Frankenstein* manuscript, notebook A.

PLATE 27 Male wax anatomical figure, Italian, 1776–1780.

PLATE 28 (*overleaf*) A lecture at the Hunterian Anatomy School, Great Windmill Street, watercolour by Robert Blemmel Schnebbelie (1781–1847).

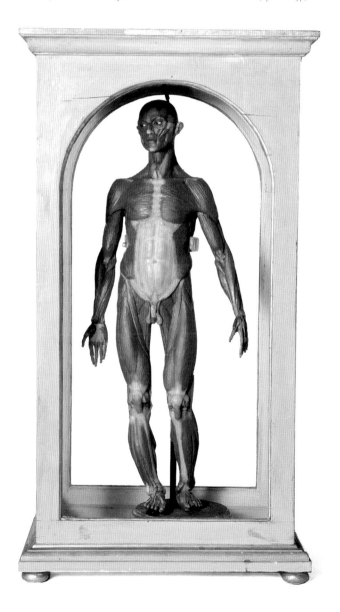

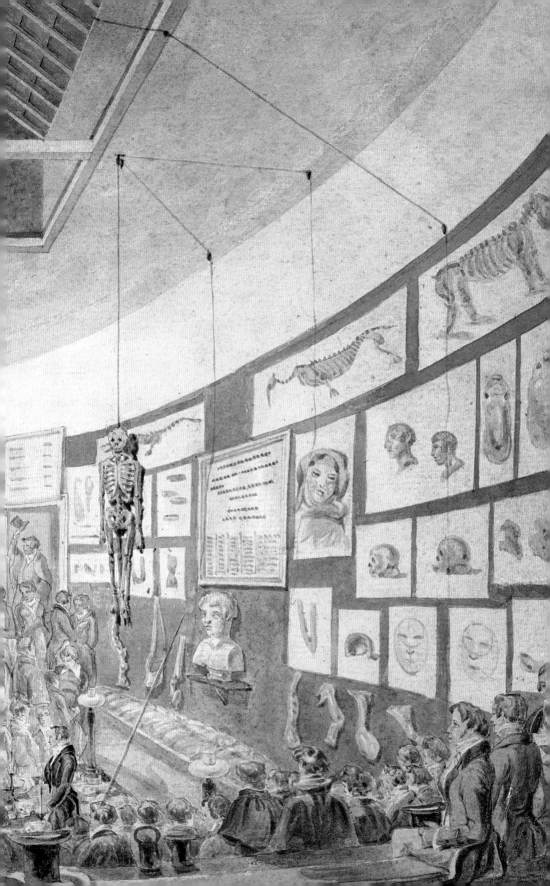

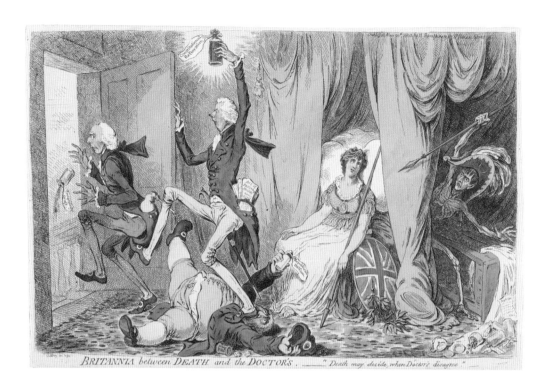

BRITANNIA between DEATH and the DOCTORS. —— "Death may decide, when Doctor's disagree."

"Hark! the Doctor knocks – she is almost done – and ready for you –" vide Old Play

BURKING POOR OLD Mrs CONSTITUTION Aged 141

PLATE 29 Contemporary political cartoon satirizing competing medical treatments. James Gillray, *Britannia between Death and the Doctor's*, etching, 1804.

PLATE 30 Wellington and Peel as the body snatchers Burke and Hare suffocating Mrs Docherty for sale to Dr Knox; representing the extinguishing by Wellington and Peel of the constitution of 1688 by Catholic emancipation. W. Heath, coloured etching, 1829.

PLATE 31 Two men placing into a sack the shrouded corpse which they have just disinterred, while Death, as a night watchman holding a lantern, grabs one of the grave robbers from behind. Thomas Rowlandson, coloured drawing, 1775.

PLATE 32 (*overleaf*) Mahogany medicine chest, c. 1820.

FIG. 14 Sir William Lawrence, mezzotint by C. Turner, 1839.

Laon and Cythna when she wrote that it is 'well that your poem was finished before this edict was issued against the imagination', but she worried that 'my pretty eclogue', *Rosalind and Helen*, 'will suffer from it'.[27] This letter demonstrates that the Shelleys were in direct contact with Lawrence at the height of his debate with Abernethy on the nature of life. Godwin knew Lawrence well too. His diary mentions 'Lawrence, surgeon' five times, including calling on him during 1815 and 1818 when the debate was at its fiercest (*see Plate 24*).[28]

JOHN HUNTER'S THEORY OF LIFE

In his first two lectures to the Royal College of Surgeons in 1814, Abernethy urged medical students to seek the truth. A higher purpose was needed, he said, to motivate and encourage students who had 'to spend our nights in study, and our days in the disgusting and health destroying avocations of the dissecting room'.[29] Victor Frankenstein fits this description well, despite the fact that he was not studying to become a surgeon. He speaks of being animated by an 'almost supernatural enthusiasm', which distracted him from the horror of his work. He tells Walton that he was forced to 'spend days and nights in vaults and charnel houses'.[30] Certainly, the work Frankenstein does destroys his health. When Henry Clerval arrives after the Creature has been animated, Frankenstein collapses, and he is dogged by ill health throughout the rest of the novel.

In his lecture, Abernethy referred to the 'great chain of living beings'. He asked, how is it possible that such different kinds of beings, organized so differently from each other, are all alive? This fact made him conclude that 'life does not depend upon organisation'.[31] John Hunter had claimed the same in his posthumously published 1794 book *A Treatise on the Blood, Inflammation, and Gunshot Wounds*: 'organization, and life, do not depend in the least upon each other'.[32]

In fact, Hunter thought that blood was a good candidate for the principle of life. He considered blood to be a living substance. In this, he seemed to have proved the truth of Leviticus, 'the life of the flesh is in the blood'.[33] Frankenstein also at times equates life with blood, commenting that the difference between him and the now-dead Justine Moritz was that 'The blood flowed freely in [his] veins'. When Elizabeth is dead, he describes her arms as 'bloodless'.[34]

There must be something, Abernethy argued, that prevents the living body undergoing 'the chemical decomposition to which dead animal and vegetable matter is so prone'.[35] There must be something that regulates the temperature of the living animal body. There must be something that causes the living body to act and move. This 'something' is found in all manner of life forms, from the simplest to the most complex. He asked, how is it that the bicep muscles of a living man can lift more than 100 pounds but that the same weight would tear the muscle if the man was dead?[36] The same physical body is capable of radically different abilities when alive or dead. Abernethy told his audience that Hunter had claimed that certain deaths (for example, by electric shock) meant that 'the principle of life may in some instances be suddenly removed …, whilst in general it is lost by degrees.'[37] What is the nature of this vital principle, he asked, if it diminishes at different speeds according to the kind of death that occurs?

These reflections made Abernethy conclude 'that irritability is the effect of some subtle, mobile, invisible substance, superadded to the evident structure of muscles, or other forms of vegetable and animal matter, as magnetism is to iron, and as electricity is to various substances with which it might be connected'. Abernethy went on to argue that electricity must be of this nature because the matter it acts upon is simply 'gross and inert'. Matter is not capable of animation on its own because it is inert, and thus 'the necessity of supposing the superaddition of some subtile and mobile substance is apparent'.[38] There was a great deal of speculation here, not least about what Hunter, now dead, had thought. In addition, Abernethy was putting forward a debatable (and controversial) notion of electricity. Even his idea that matter was inert, which meant that the vital principle would be found elsewhere, was up for debate.

Abernethy continued to find further points of comparison between electricity and the vital principle. Electricity formed an 'important link in the connexion of our knowledge of dead and living matter', he argued, and cited Humphry Davy to prove this. Indeed, he considered that Davy had – by means of electricity – been able to 'control the ordinary operations of nature'.[39] Davy certainly had used exactly this language of mastery and control in his lectures and publications. We see this language in Professor Waldman's and Victor Frankenstein's words in the novel. It betrays the masculine view of nature as female, submissive and to be conquered. For Abernethy, electricity was the means by which to control nature.

Abernethy claimed that he had always been sure that electricity had the property of being a 'something', or a substance: 'That electricity is something, I could never doubt, and therefore it follows as a consequence in my opinion, ... that it enters into the composition of every thing, inanimate or animate.' Somewhat bizarrely, he backtracked from stating that electricity was the vital principle: 'It is not meant to be affirmed that electricity is life.'[40] In fact, this seemed to be exactly what Abernethy was arguing. At the very least, he was saying that electricity was analogous to the vital principle, if not that they were one and the same thing. Abernethy compared it to the spirit that had been known as the *anima mundi*, or soul of the world.

Though Abernethy was not brave enough to state explicitly that electricity was the principle of life, it was clear to his audience that this was what he thought. Hiding behind Hunter's supposed 'theory of life' was one way to avoid having to state this as his own theory. But there were so many questions left unanswered. Did electricity work in the way that Abernethy described? Was the matter of the physical body so inert that it required the addition of something else to enable life? Were men of science any closer to discovering what life is?

FRANKENSTEIN AND ABERNETHY

Frankenstein frames his research question in the same way as Abernethy: 'One of the phaenonema which had peculiarly attracted my attention was the structure of the human frame, and, indeed, any animal endued with life.'[41] Here, he contemplates the 'structure of the human frame', by which he means the body, or the physical organization of the living being. Given that living animals come in such different kinds of bodies, what is it that gives them life? Comparative anatomy, a discipline at which Hunter excelled, showed many similarities between living creatures: the fin of a fish functioned in the same way as the wing of a bird, for example. But when you extended your view of living creatures to molluscs, Venus flytraps and eggs, what did they all have in common? They do not all possess digestive organs, nervous systems or even brains. They do conduct electricity, though, and this fact had led Luigi Galvani and others to believe that all animals possessed a peculiar kind of electricity (*see Chapter 3*).

Frankenstein certainly thinks that electricity is 'something', or that it has a material substance. He calls it a fluid in the novel.[42] The verbs used demonstrate that 'something' has been added to the body. For example, the Creature is 'endowed' with 'perceptions and passions'.[43] Life is 'given' by Frankenstein. He tells Walton that he worked for two years with the 'sole purpose of infusing life into an inanimate body'.[44]

The Creature is not grown. Life does not emerge or come into being gradually. It is, instead, instantaneous and immediate, the result of 'something' being introduced to an otherwise dead and inert body.

When Frankenstein describes the churchyard as 'merely the receptacle of bodies deprived of life', he suggests that life is something that can be given and taken away. It is added to the body, not an inherent product of it. Eventually, after years of study, Frankenstein is able to animate a body in the same way that Abernethy believed it was achieved: 'I became myself capable of bestowing animation upon lifeless matter.'[45] Matter is lifeless, therefore, before animation is 'bestowed' upon it. Frankenstein sets about preparing a body for the reception of the vital principle and this poses its own difficulties: 'Although I possessed the capacity of bestowing animation, yet to prepare a frame for the reception of it, with all its intricacies of fibres, muscles, and veins' was difficult.[46] There is a clear division of labour here. Having worked out what the vital principle is and how to infuse it into a body, he then has to create a body to receive it. The practice set out here fits Abernethy's theory entirely. The body is merely a frame, a receptacle, for animation to be bestowed upon it.

THE MORAL CONSEQUENCES

At the end of his 1814 lecture, Abernethy announced that 'The contemplation of this subject at large, is fitter for meditation in the closet than for discussion in the lecture-room', using closet in the older sense of a private or secluded room in which one might pray or reflect.[47] The closet was also the room where you might dress or have a private conversation. For similar reasons, in the late twentieth century the 'closet' became a metaphorical space in which people would conceal their sexuality from the world. Closet drama, such as Percy Shelley and Lord Byron produced in *Prometheus Unbound* and

Manfred, was called such because it was not intended for the wide audience of a public stage.

When Abernethy said that the vital principle was a subject for the closet, he meant that it was a subject that had moral, religious and personal implications. Those who see life as the same in all living creatures, from humans to animalcules, would not be able to claim that there is anything especially moral or divine about humans, and this idea is sacrilegious. The precise implications of Abernethy's theory become clear at the end of his lecture. If you believe what he says about the vital principle, then you will also believe 'that in addition to his bodily frame, [man] possesses a sensitive, intelligent, and independent mind: an opinion which tends in an eminent degree to produce virtuous, honorable, and useful actions.'[48]

The stakes were high. Accepting Abernethy's theory of life meant that you also accepted that there was more to human life than just the physical body. Good people were the product of the vital principle that Abernethy advocated in his lectures. His comment shows that medical lectures strayed far beyond what might be covered today, with our contemporary sense of medical objectivity and professionalism. Disagreeing with Abernethy meant taking on the medical establishment.

XAVIER BICHAT AND THE FRENCH MATERIALISTS

At the end of the eighteenth century, the French anatomist Marie François Xavier Bichat had famously defined life as 'the sum of the functions, by which death is resisted'.[49] Bichat was known for the number of dissections he undertook as well as his experiments on living and dead animals. He notes in his *Physiological Researches on Life and Death* that while the nature of life was not known, it could be witnessed in its effects. Life was to be measured by the difference between 'the effort of exterior power' and that of 'interior resistance';[50]

in other words, the degree to which the ordinary physical pressures of the world (such as gravity) were resisted by this mysterious inner life force. He pointed out that life in a child is in abundance. In an adult there is a balance between outer and inner force. In an old person outside pressure continues as it always has done but the inner force's ability to resist it diminishes.

Bichat makes an important distinction between vegetable and animal life. Vegetable life, he argues, is made up of a succession of assimilation and excretion. Plants continually absorb what they need and release that which they no longer need. For example, plants take in and transform air, water and sunshine into their own nature. Then they release the substances which are not of their nature, such as oxygen. They live in one spot, rooted to the earth. By comparison, animals are free to roam as 'an inhabitant of the world'.[51] Not only this, animals can feel, perceive, reflect upon their sensations, voice desires, fears, pleasures and pains. Identifying these fundamental differences, Bichat concludes that there are two kinds of life: organic (vegetable) and animal (including human life). In his 1814 lecture, Abernethy described Bichat as the 'industrious and ingenious French anatomist' and his division of life is noted with approbation. Before long, though, using Bichat's name would become a red flag to those supporting Abernethy against Lawrence.

WILLIAM LAWRENCE'S *INTRODUCTION*

On 21 and 22 March 1816 William Lawrence took his turn to ascend the stage at the Royal College of Surgeons and deliver his first two lectures as Professor of Anatomy and Surgery. These were publicized as introductory lectures to medical students on the subjects of Comparative Anatomy and Physiology. When these lectures were published in July the same year, the title page noted some of

Lawrence's many accolades. He made a point of expressing the 'sentiments of respect and gratitude' that he held for Abernethy.[52] He noted that because he had 'lived for many years under his roof', he was able to speak of him as a 'Man and a Friend'.[53] Despite this, Lawrence took a very different approach to the question of what life is.

Lawrence emphasized the power that all living creatures had – from the hairworm (*Gordius*) to newts, crabs and humans – to repair themselves when 'bent, broken, worn, or spoiled'. He shared horrid experiments, conducted by others, in which tortoises survived having their brain removed, newts regrew their tails and crabs their claws, and the wheel animal (*Vorticella rotatoria*) revived when moisture was applied to it after years left dried and seemingly dead. While it seems that humans have less of this simple yet powerful vitality, he pointed out that we are able to knit together the skin of a wound, repair broken bones, and accomplish 'restoration'.[54] Lawrence is convinced that the dissection of animals has been of crucial importance in understanding the digestive, respiratory and generative processes that humans perform. Frankenstein, of course, also 'tortured the living animal to animate the lifeless clay' as part of his research.[55] One major problem in the study of human life is that subjects are usually dead (rather than living) when they are studied.

Lawrence's first lecture offered a history of comparative anatomy, starting with Aristotle and reaching his present day. The lecture displayed his education and knowledge, from classical sources to an impressive array of European texts. At times, the notes in the published version of the lectures take up more space on the page than the text of the lecture itself. He certainly did not confine himself to, nor privilege, British men in his account. Compared to Abernethy's lectures, Lawrence's were vastly more cosmopolitan and scholarly. Lawrence even criticized Britain for not having the proper resources for the study of comparative anatomy when France had this in the form of the Jardin des Plantes in Paris. Lawrence exclaimed that, in Britain, 'We have no national collection of living animals, no museum of natural history, no public institution for teaching natural science.'[56] Lawrence also criticized the universities in England (of which there were only two at the time, Cambridge and Oxford) for not teaching natural science. He claimed that the results of these failures was that 'Zoological pursuits have languished in England during [the] great part of the past century.'[57] He was already skating on thin ice, denigrating his native country and finding it wanting in comparison to the very country it had, less than a year before, bested in war. The Battle of Waterloo had ended decades of fighting between the British and French. Lawrence's superiors would not want to hear that the French were more advanced in any respect. (*Plates 20, 21, 29 & 30* show how medicine was employed in political cartoons.)

In Percy Shelley's 'Preface' to *Frankenstein*, he mentions 'Dr. Darwin, and some of the physiological writers of Germany' as possible sources for the idea that a being could be animated.[58] The University of Ingolstadt, which Frankenstein attends, is in Germany and Percy may have been thinking of Blumenbach, whom Lawrence had translated in 1807. Lawrence reserved much praise in his lecture for the achievements of comparative anatomists in Germany. He added more names to Blumenbach's, including Samuel Thomas von Sömmering, Johann Abraham Albers, Ludolph Christian Treviranus, and others. Intriguingly, though, when Frankenstein thinks about creating a mate for the Creature, he claims that he needed to visit London:

> I found that I could not compose a female without again devoting several months to profound study and laborious disquisition. I had heard of some discoveries having been made by an English philosopher, the knowledge of which was material to my success.[59]

Who was this individual that he needed to visit? Did Mary Shelley have someone in mind? When he arrives in London, Frankenstein tells Walton that he 'quickly availed myself of the letters of introduction that I had brought with me, addressed to the most distinguished natural philosophers'.[60]

In his lecture, Lawrence argued that the study of medicine 'has the strongest claim on the attention of every liberal man'. Like Abernethy he asserted that it 'exerts a beneficial and important influence on the moral dispositions'. Perhaps differently to Abernethy's idea of morality, Lawrence's use of 'liberal' suggests that he is thinking of someone who is open-minded and tolerant in his politics. He described medicine as a 'tranquil occupation' for the mind and a 'salutary contrast to the restless agitation of avarice and ambition'. It is a source of 'innocent pleasures', according to Lawrence, 'well calculated to detach us from the frivolous and destructive pursuits of dissipation or debauchery'.[61] Certainly Frankenstein is brought up in a particularly liberal environment in Geneva, in the tradition of Jean-Jacques Rousseau, and educated well by his civic-minded father Alphonse. He does not evince the seductions of money and greed. But 'lofty ambition' is one of his motivations and a failing since childhood.[62] The disregard for ethics in Frankenstein's project, though, is one of the most serious criticisms that can be levelled against it. His moral failings lead him neither to question whether the project is right, nor to take responsibility for the life that he creates.

LAWRENCE ON LIFE

Lawrence opened his second lecture in 1816, on 'Life', in a controversial manner. Anatomy and physiology have to be taken together, he argued. What would be the point in having the working of a watch or a steam engine described without reference to their function? The 'science of organization' should not be separated from the science of 'life', he claimed. He supported this with further explication: 'organization is the instrument, vital properties the acting power, function the mode of action, and life the result'.[63] In short, life is the result of organization. Lawrence stated the exact opposite to what his mentor Abernethy had asserted just two years previously.

Lawrence also noted, in a way clearly influenced by Bichat, that the living body resists the physical forces at work upon it. When the body dies, it is no longer able to resist these forces. And he went into detail about the changes that take place upon death. Frankenstein also studies death and takes note of the changes that occur: 'I beheld the corruption of death succeed to the blooming cheek of life.' Elizabeth similarly notes the change that takes place in William Frankenstein: 'About five in the morning I discovered my lovely boy, whom the night before I had seen blooming and active in health, stretched on the grass livid and motionless.'[64] Lawrence's explanation for such alterations was that these 'are the effects produced by the chemical action of the solids and fluids upon each other, and by the affinities of the surrounding agents air, moisture and heat, to both.'[65] Whereas during life the 'vital forces were superior to these chemical affinities, and superseded their action', after death, the universe reasserts its power over the body, which becomes subject to the usual laws of physics and chemistry.[66]

What becomes apparent in Lawrence's and Bichat's texts is that life is the surprising and unusual circumstance. The body far more naturally tends towards a state of death, disintegration and decay. Lawrence asked how it is that the living human body can keep a standard bodily temperature in situations that would normally freeze or boil the matter of which they are made. He noted that the living body maintained the animal heat necessary for life 'in the intense colds' of Spitzbergen and Greenland, a fact to which Captain Walton, his ship's crew, the Creature and Victor Frankenstein attest.[67]

Lawrence emphasized, as Bichat had, the constant absorption and elimination of materials that the body needed in order to live. The living body, as he described it, is in constant motion: 'we see a continued change, so that the body cannot be called the same in any two successive instants.'[68] There is a constant circulation of materials, new replacing old continually, leading to the constant renewal of life.

Towards the end of his lecture, Lawrence opposed his mentor Abernethy explicitly. Lawrence said that Hunter 'did not attempt to explain life … by the illusory analogies of other sciences', with the suggestion being that Abernethy has done exactly this. Invoking *Hamlet*, Lawrence's target became unmistakeable and his ridicule clear: 'this vital principle is compared to magnetism, to electricity, and to galvanism; or it is roundly stated to be oxygen. 'Tis like a camel, or like a whale, or like what you please.'[69] It was a matter of some debate whether, as Abernethy claimed, electricity was in fact a fine fluid, superadded to the material on which its effects were witnessed. Lawrence shed doubt on everything that Abernethy had argued for in his lectures a year ago. And he was unequivocal: 'The truth is, there is no resemblance, no analogy between electricity and life.'[70] As if enough damage had not already been done, Lawrence concluded with a statement that was rude and ill-mannered:

> It seems to me that this hypothesis or fiction of a subtle invisible matter … is only an example of that propensity in the human mind, which has led men at all times to account for those phenomena, of which the causes are not obvious, by the mysterious aid of higher and imaginary beings.[71]

Believing in Abernethy's idea of life was tantamount, Lawrence argued, to believing that supernatural beings control our fate. He flew dangerously close to saying that believing in God is for idiots. He ended by quoting Lucretius' *De rerum natura*, as if the associations with materialism were not clear enough.[72] Lawrence was playing with fire. Britain at this time was a conservative, religious and traditional country. The Shelleys had first-hand experience of the treatment and exile that non-conforming individuals, who flouted marriage, would encounter. Materialism, atheism and radicalism would not be tolerated.

ABERNETHY'S *PHYSIOLOGICAL LECTURES*

Abernethy could not take this insubordination lying down. He returned to the Royal College of Surgeons to give another lecture the year after Lawrence had spoken there. He recapitulated the controversial part of his first lectures, that electricity was analogous to life, asserting that 'my meaning has been either misunderstood or misrepresented'.[73] Throughout, he continued to say that he was presenting Hunter's theory of life, rather than his own. This made disagreeing with it tantamount to disagreeing with Hunter. But he did, reluctantly, add his own personal opinion too. He claimed that there was a 'subtile substance belonging to living bodies, a principle of life'. Abernethy felt that in these lectures he had to defend himself against the criticism of a certain party of gentlemen, whom he called 'the Modern Sceptics'. There was a serious difference of opinion: 'I discover that they wish me to consider life to be nothing.'[74] The idea that life was nothing was the opposite of Abernethy's conviction that there was a vital principle, a 'something', which animated otherwise dead matter. There was more riding on this terminology though. To consider life to be 'nothing' was to be a nihilist, a radical. It implied that life is insignificant, without value or use, trifling and unimportant. It went against Abernethy's theory of life, which creates virtuous, honourable and useful people. It smacked of French materialism and was a dangerous view to spread among young, susceptible medical students.

Abernethy asked what motivated the ridicule he had been subjected to in Lawrence's lecture. Is it, he asked, that these sceptics cannot accept that there is 'any thing which is not an object of sense'?[75] Does a thing need to be seen or demonstrated in order to be believed? The implications of his questions are perilous: if this were the case, we would not believe in the soul or in God. Quoting *Macbeth*, he countered Lawrence's *Hamlet*. He asked whether the Modern Sceptics' insistence that life is the product of organization was in fact 'because they wish to persuade others … "that when the brains are out, the man is dead"'.[76]

Abernethy was pleased to boast that the theory he put forward was 'productive of nothing but good to humanity', suggesting that Lawrence's theory was morally dubious. Belief in the 'distinct and independent nature of mind, incites us to act rightly from principle', Abernethy claimed. He was not surprised that the French held such degenerate views of life when they were degenerate themselves. He described France as a 'nation where the writings both of its philosophers and wits have greatly contributed to demoralize the people'. The English, in contrast, are a 'thinking people, who consider the probable ends of conduct from its beginning'. He declared that considering the moral consequences of an action was what characterized the English. This is something that Frankenstein patently does not do when he animates the Creature, but then he was French-speaking Swiss. Lawrence's words and the French materialists by which they were influenced had a 'pernicious tendency' according to Abernethy.[77]

WHAT HAPPENED NEXT

After the publication of *Frankenstein*, Lawrence gave one more devastating and unflinching lecture to the Royal College of Surgeons in 1819, published as *Lectures on Physiology, Zoology, and the Natural History of Man*.[78] His punishment came swiftly: he was suspended from his post as surgeon at Bridewell and Bethlem hospitals. Although he said that he would never be silenced, Lawrence quickly wrote a letter recanting his opinions and promising to suppress his publications. His 1819 lectures were withdrawn within a month of publication.

There were a number of attacks from eminent men. Thomas Rennell, the Christian Advocate at the University of Cambridge, published *Remarks on Scepticism*, which attacked both Bichat and Lawrence. Surely, Rennell asked, Lawrence cannot be 'desirous of infecting the unwary youth committed to his charge, with principles subversive of all private happiness, all social morality'?[79] Edward William Grinfield, a minister in Bath, warned against 'the evils which [Lawrence's views] portend to society in general, and to the morals of your own profession in particular'.[80] The *Quarterly Review* claimed that the superadded vital principle which Abernethy espoused had been 'added by the will of Omnipotence'.[81] Lawrence's statement in his 1819 lectures that 'An immaterial and spiritual being could not have been discovered amid the blood and filth of the dissecting room' went unheard.[82] Religion definitely won this battle with science.

Lawrence bought up remaining copies of his offending *Lectures on Physiology*, but the book flourished in America and pirate editions soon came out. He tried to obtain copyright (and some of the profit) for the sale of these editions, in a Court of Chancery case heard in March 1822. But the counsel for James Smith, one of the pirate publishers, told Lord Eldon that Lawrence's lectures taught that 'a man had no more soul than an oyster, or any other fish or insect'.[83] After the Lord

Chancellor ruled that it was unsuitable for public reading, ironically, radical publishers were free to continue to disseminate as many cheap and accessible pirated versions of Lawrence's *Lectures* as they wished. Southey's *Wat Tyler*, Percy Shelley's *Queen Mab* and Byron's *Cain* had suffered the same fate. When Lawrence admitted to giving some 400 copies to medical readers, he was suspended again from his hospital posts. Once again, he recanted his theory of life and the radical journal *Monthly Magazine* published his letter in parallel columns alongside Galileo's retraction, making its position on the debate clear.[84] He was allowed to return to his posts and eventually, having learned his lesson, would become known for his antipathy to reform. By 1837 he was thoroughly an establishment man, named Surgeon Extraordinaire to the new queen, Victoria, and later her Sergeant-Surgeon.

The question of whether the Creature in *Frankenstein* has a soul has been asked many times. The term is used a number of times in the novel, often to denote poetic or elevated thoughts and sensations. For critics of Lawrence, the soul was either analogous to the vital principle or the vital principle itself. At the very least, the existence of a soul demonstrated that something too subtle for the physical senses to perceive could be superadded to the inert matter of the body. When Safie arrives, the soul is designated – metaphorically – as the receptacle of animation: 'The arrival of the Arabian now infused new life into [Felix's] soul.'[85] The debate between Abernethy and Lawrence demonstrates how polarized opinions on the nature of life were and just how controversial this issue was. But this was not the only controversial aspect of the study of medicine at this time. It was known that the bodies used in anatomy-lecture demonstrations were obtained illegally, stolen from graves and sold to surgeons and physicians. Mary Shelley's novel would have reminded readers that their own corpses were not safe from an anatomist's dissection and medical display after death.

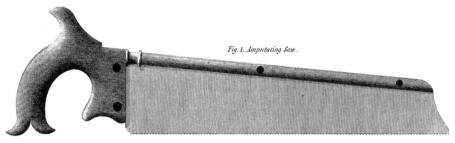
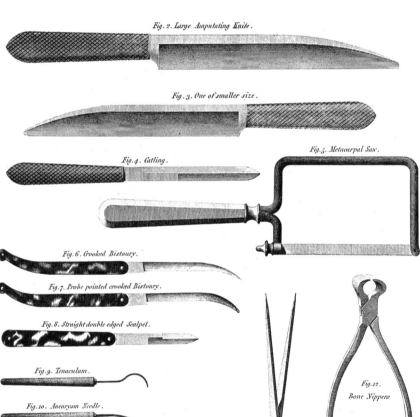

FIVE
RAISING THE DEAD

The night Victor Frankenstein animates the Creature, he has a terrible nightmare. Fraught with the horror of his creation, Frankenstein dreams that he meets his beloved, Elizabeth, in the streets of Ingolstadt. When he kisses her lips 'they became livid with the hue of death'. Her features also change. Instead of his fiancée, he finds that he is holding the 'corpse of my dead mother in my arms; a shroud enveloped her form, and I saw the grave-worms crawling in the folds of the flannel'.[1] He wakes up, horrified; he is sweating, his teeth are chattering, and his limbs convulse. He realizes that the Creature is watching him, and at this moment Frankenstein runs away. The episode presents the transformation from life to death. It also dwells on the horror of contemplating the reality of the death of our loved ones.

Previously, when Frankenstein had described how he discovered the principle of life, he boasted that he did not fear the dead. He tells Captain Walton that his father had been careful to ensure that his education was rational and that 'supernatural horrors' should not frighten him. As a result, Frankenstein considered that 'a churchyard was … merely the receptacle of bodies deprived of life, which, from being the seat of beauty and strength, had become food for the

FIG. 15 Set of English surgical instruments, John Weiss & Sons, c.1830.

worm.'[2] There is nothing Gothic in this view of death. Death is merely the absence of life. Indeed, dead bodies are often labelled merely 'lifeless' in the novel.[3] Frankenstein is unsentimental, perhaps showing scientific objectivity, when he insists on mentioning the worms that feed on dead bodies. This objectivity is absent from his nightmare, though, when he imagines 'grave-worms' on his mother's corpse. In the seventeenth century, William Harvey, famous for discovering how the blood circulated, privately dissected both his father and his sister. It is unlikely that Frankenstein would have been capable of the same.

Frankenstein recalls that when he was young, and impressed by Albertus Magnus, Paracelsus and other alchemists, he tried to raise ghosts using the incantations found in their ancient books. At this point he attributed his lack of success to his own inadequacies rather than thinking that there were no spirits to call. There is a marked difference in the older Frankenstein, whose beliefs about the world are changed after witnessing 'natural phenomena' such as distillation, the effects of steam, the air pump, and lightning destroying a tree. The older Frankenstein is no longer frightened by ghost stories. Darkness is nothing to him. He is a rational man, intent upon looking at everything in the world dispassionately. But this professionalism ends when he animates the Creature, by whom he is repulsed and horrified.

The dead are crucial to Frankenstein's research. He tells Walton that 'To examine the causes of life, we must first have recourse to death.'[4] It is certainly difficult – if not impossible – to study life while the subject is living. But here Frankenstein seems to be saying that there is a connection between living and dead matter, such that they help to explain each other. In order to discover what life is, he needs to 'observe the natural decay and corruption of the human body'. He spends 'days and nights in vaults and charnel houses', examining the effects of death upon the body. Thus death illuminates his work:

'I aw how the fine form of man was degraded and wasted; I beheld the corruption of death succeed to the blooming cheek of life; I saw how the worm inherited the wonders of the eye and brain.'[5] It is precisely at this moment that Frankenstein discovers the secret of life.

THE GUILLOTINE

As discussed in Chapter 2, hanging was not at this time an effective way to execute people. In 1789, the anatomy professor Joseph-Ignace Guillotin addressed the French National Assembly to argue that it was unfair that aristocrats were beheaded by the sword, a more effective death, while commoners had to suffer more through other, less reliable methods. In 1792 trials began of the killing machine that would bear his name, the 'guillotine'. The effectiveness of this method exceeded anything that Guillotin could have imagined. It enabled the mass killing of thousands in quick succession. People could be beheaded at a rate of one per minute. At the height of its use, 300 people were guillotined in only three days. During the Reign of Terror, from 1793 to 1794, around 1,700 persons were executed in this manner.

The Terror provided French anatomists with far more bodies than their British counterparts had access to and may well have enabled French medicine's great advancements in the nineteenth century. Allegedly, the anatomist Marie François Xavier Bichat dissected over 600 decapitated bodies within six months of the French Revolution. The guillotine was proposed as a more humane way to end a person's life and was represented as offering an instantaneous death, without pain or unnecessary suffering. But its use raised the question of whether the vital principle inhabited the brain or, say, the heart. It tested the proposition that the vital principle remained in the body for some time after death, as was argued by those who thought it was electricity (*see Chapter 3*). It offered graphic evidence to answer the question of whether death occurred at the very instant the head was separated from the body.

A very famous case seemed to suggest that life continued for a few seconds after a person was beheaded. On 17 July 1793, Charlotte Corday, sentenced to death for stabbing the Jacobin leader Jean-Paul Marat in his bath, was guillotined. After she had been beheaded, a carpenter called Legros held up her head and slapped her face. Onlookers asserted that a blush spread across her cheek signifying a look of indignation. The blush seemed to show Corday's continued awareness of the outrage she endured. If this was the case, it meant that consciousness continued in the body after the separation of the head. A number of eminent physicians claimed that, for at least a few moments after death, feeling and identity lingered. The guillotine's use further fuelled debate over where life ended and death began.

It seemed ridiculous to the lay person that there should be any confusion between the states of life and death. Reviewing the Royal College of Surgeons debate between John Abernethy and William Lawrence (*see Chapter 4*) in 1814, an anonymous *Edinburgh Review* article was incredulous: 'physiologists are not yet agreed as to the precise grounds even of that most familiar of all classifications – the arrangement of Bodies into *Living* and *Dead*; ... there is not, at this moment, a term which is used with greater ambiguity, than the term Life.'[6] It should be perfectly clear, the reviewer implies, when a person is alive or dead. Yet, at every turn in this period it was acknowledged that the issue was far more complicated than might have been thought. Mary Shelley capitalizes on this uncertainty in her novel.

FROM DEATH TO LIFE

Frankenstein's achievement in creating a living, breathing being is only possible because he thinks that the boundary between life and death, and vice versa, can be transgressed. When he realizes that he possesses the secret of life, he proclaims: 'Life and death appeared to me ideal bounds.'[7] He means 'ideal' in the sense that they exist only as an idea confined to thought or the imagination, as opposed to being material or actual. The boundary between life and death is thus abstract rather than real. Considering the division between life and death this way is what enables him to move between them. There are a number of episodes in the novel where characters exclaim upon how quickly death succeeds life, or how swiftly a living body becomes a corpse.

Crucially, though, Frankenstein is also fascinated by the other side of this coin, the way that death succeeds life. While the individual may be lost forever, on a grander scale life and death exist in a cycle of transformation where the dead provide for new life. Frankenstein brought life to something that was previously dead: he found himself able to 'bestow animation upon lifeless matter'. He initially hoped that he would be able to 'renew life where death had apparently devoted the body to corruption' and is adamant that he will be able to 'break through' the boundary that divides life from death.[8]

Frankenstein seems to believe that life and death exist on a continuum, rather than as strictly separate states of being, and that studying one can lead to the discovery of the secrets of the other. He describes to Captain Walton how he set about 'examining and analysing all the minutiae of causation, as exemplified in the change from life to death, and death to life'.[9] He does not merely see this as a one-way street, where life leads inexorably to death, but is also interested in the journey from death to life. When he notes that dead bodies provide 'food for the worm', he is alerting us to the fact that the dead provide for future life in a cycle of existence. For example, manure provides for the life of plants and vegetables. As we have now confirmed, after death the remains of the body are broken into their biochemical parts, re-enter the physical world and become subsumed again into the cycle of life.

From 1808 to 1825 eminent chemists, physiologists and anatomists formed the Animal Chemistry Society, with the express aim of studying life. We would now call the field that these men were bringing into being 'organic chemistry'. The group had a clear political and moral character. They were all vitalists like John Abernethy and John Hunter. While they investigated living and dead matter, they were also certain that both were subservient to a vital principle. As Humphry Davy, one of the members of this society put it,

> The laws of living and dead nature appear to be perfectly distinct: material powers are made subservient to the purposes of life, and the elements of matter are newly arranged in living organs; but they are merely the instruments of a superior principle.[10]

In other words, the Animal Chemistry Society believed that life could not be explained by or reduced to chemistry. There was something else, beyond matter, at work too. Their research, which had the backing of the Royal Society, can be seen as providing more evidence for the conservative, vitalist view of life that triumphed in the debate at the Royal College of Surgeons, discussed in the previous chapter.

Members of the Animal Chemistry Society met in each other's houses to discuss their investigations, and the Royal Society's journal, *Philosophical Transactions*, had first refusal when their papers were published. The journal disseminated Society members' work on 'Animal Fluids' and secretions, including blood and urine; the 'formation of fat in the Intestines'; 'the influence of Nerves' and 'the secretions of the Stomach'; 'the different Modes in which Death is produced by certain vegetable Poisons'; and 'the influence of the Brain in the generation of Animal Heat'.[11] The Society's members experimented with a wide range of methods, including decomposing substances with electrolysis or using their knowledge of blood circulation to enhance the effective operation of medicines. Frankenstein's work would have fitted well with their aims and objectives.

Frankenstein notes four times in the novel that worms benefit from the decay of once-living matter, including his statement that he 'saw', first hand, 'how the worm inherited the wonders of the eye and brain'. Davy's *Elements of Agricultural Chemistry*, which Percy Shelley wrote extensive notes on in 1820, also states that decomposed animal matter is essential to the life of those lower down the chain of being, and that this vegetable life is – in turn – essential for living animals. This is the cycle of life at its most obvious. Percy notes specifically: 'Manure is useful & may be converted into organized bodies.' Speaking about the decaying animal body, Percy's notes sum up Davy's writings with: 'from the progress of waste of animal life a principle necessary to the existence of vegetables is produced'.[12] Frankenstein equally knows that death creates life just as life creates death.

LIFE AS STIMULATION

Initially, Victor Frankenstein intended that his research would revoke death rather than create a new life. He considered the possibility that he might be able to make people immortal. He momentarily imagines, in a conceited way, 'what glory would attend the discovery, if I could banish disease from the human frame, and render man invulnerable to any but a violent death!'[13] As discussed in Chapter 2, the violent deaths featured or mentioned in the novel include strangulation (the Creature's method of killing), hanging (presumably how Justine was killed) and drowning (a fate Walton's crew face throughout the novel and which Frankenstein himself nearly dies from). Frankenstein's father's death, which comes late in the novel, is brought on by these violent deaths: 'He could not live under the horrors that were accumulated around him; an apoplectic fit was brought on, and in a few days he died in my arms.'[14] The phrase 'apoplectic fit' might indicate a brain haemorrhage. Other non-violent deaths would include those caused by disease. For example, Frankenstein's mother dies of scarlet fever, which she catches from Elizabeth. His mother's death, coming at an impressionable age, recalled by him in the narrative of his life to Walton, may well be a key motivation for his studies.

The prevalence of illness and disease seemed to suggest that *'Life is a forced state'* as John Brown, a Scottish physician, argued.[15] In his *Elements of Medicine*, Brown explains further: 'To every animated being is allotted a certain portion only of the quality or principle, on which the phenomena of life depend. This principle is denominated EXCITABILITY.'[16] In Brown's system, all diseases are classified as resulting either from too much excitability (sthenic) or too little excitability (asthenic). If the latter disease was diagnosed, a stimulant – usually opium – was given to restore the balance of the nervous system. Brown himself was known for enlivening his lectures at the University of Edinburgh by self-administering laudanum (opium dissolved in whisky). Allegedly Brown had a thermometer with which to measure the degree of excitability in a person's body. His system was briefly very influential in Britain, but even more so on the Continent. We can see Brown's language in some important Romantic-period writing, notably in Wordsworth's Preface to the *Lyrical Ballads*, where he criticizes modern lifestyles that encourage a 'degrading thirst after outrageous stimulation'.[17] In *Biographia Literaria* (1817), the opium user S.T. Coleridge accuses Charles Maturin's play *Bertram* of making the reader depraved and 'craving alone the grossest and most outrageous stimulants' in literature.[18] *Frankenstein* shows just how commonplace the use of opium was. Victor tells Walton: 'Ever since my recovery from the fever I had been in the custom of taking every night a small quantity of laudanum.' When he has trouble sleeping after Henry Clerval's death, Frankenstein 'took a double dose' of laudanum and is plagued by nightmares.[19] There is much evidence in the novel that Mary Shelley thought about health and disease in precisely the way that Brown taught.

Frankenstein is often unwell and the state of his health has been linked to the Creature's deficiencies. For example, Erasmus Darwin, whom Mary refers to in her 1831 introduction to the novel, was among those who believed that the mind of the father determined the characteristics of the unborn child.[20] Mary Shelley is specific about Frankenstein's illness. During the months preceding the creation, he describes how the usual excitement that he experiences in life has been replaced by another kind of excitement. He is 'animated by an almost supernatural enthusiasm'.[21] He diagnoses his illness just before that 'dreary night of November' on which he creates the Creature:

> Every night I was oppressed by a slow fever, and I became nervous to a most painful degree; a disease that I regretted the more because I had hitherto enjoyed most excellent health, and had always boasted of the firmness of my nerves. But I believed that exercise and amusement would soon drive away such symptoms; and I promised myself both of these, when my creation should be complete.[22]

He has a nervous illness, defined using a Brunonian pathology in which illness is caused by or results in unnatural excitement. The word 'enthusiasm', which he uses, has added connotations of madness. Frankenstein is driven by a sole, monomaniacal pursuit, which he describes as an 'unnatural stimulus', using a term distinctly associated with Brown.[23] After giving the Creature life, Frankenstein's illness reaches a crisis and he falls 'down in a fit'. Later, after seeing Clerval's body, Frankenstein again experiences 'strong convulsions'. He succumbs to fever and raves like a madman. The 'nervous fever' persists on and off through the novel until his end.[24]

Elsewhere in the novel, other illnesses are diagnosed according to Brown's definitions. When Elizabeth describes the death of Justine's mother, for example, it is in these terms: 'Perpetual fretting at length threw Madame Moritz into a decline, which at first increased her irritability, but she is now at peace for ever.'[25] These are similar symptoms to those witnessed in Frankenstein's illness. These symptoms are represented as having been an essential aid to Frankenstein's work in animating the Creature but they become immediately debilitating once he is alive. Likewise, Madame Moritz's illness first increased her life force, or 'irritability', but then exhausted it. In Brown's model, the excessive excitement that the illnesses of both Victor and Justine's mother perpetuate eventually consumes them.

The death of Frankenstein affords another opportunity to examine how Mary Shelley represents life in the novel. He seems to die of the same nervous disorder that first afflicted him when he was occupied by the creation of the Creature, although there is always the potential that this nervous illness is also, largely, a mental disorder. Frankenstein tells Walton, for example, that the 'sight of a chemical instrument would renew all the agony of [the] nervous symptoms' of his original illness.[26] Walton writes to his sister that 'a feverish fire still glimmers in [Frankenstein's] eyes; but he is exhausted, and, when suddenly roused to any exertion, he speedily sinks again into apparent lifelessness.' At the end of the novel Frankenstein faints, having used his last reserve of energy to encourage, as Ulysses fatally did, the ship's crew to continue Walton's foolhardy mission in the name of glory. His collapse is described by Walton as having the appearance of death: 'It was long before he was restored; and I often thought that life was entirely extinct.' When Frankenstein's death does come, it is reported in matter-of-fact terms, with Walton's elegiac mourning of the 'extinction of this glorious spirit' interrupted by the Creature's appearance.[27] The scene in which Walton meets him incorporates Frankenstein's dead body and there are references and gestures to the corpse during their conversation. The Creature wonders if and how Frankenstein now exists, questioning whether, 'if yet, in some mode unknown to me, thou hast not yet ceased to think and feel'.[28] The way this question is framed suggests the possibility of a fundamental difference between the speaker and his creator. It may be a 'mode unknown' to the Creature because he has not experienced death or because he thinks that his death will not be the same. These words may suggest that the Creature does not have a soul and thus his afterlife will be different.

THE RIGHTS OF THE CORPSE

For religious reasons, proper burial mattered a good deal to people at this time. This was partly why the provision in law had been made in King Henry VIII's reign that murderers' bodies could be dissected. The idea that this punishment would 'better prevent' the crime was made explicit in the Murder Act of 1752, which insisted that 'in no case whatsoever shall the body of any murderer be suffered to be buried'. Being denied a grave, and being cut up into many pieces, raised problems for people who believed that the body needed to be intact and in one place in the event of Judgement Day. When the dead are resurrected, how would the dismembered body be made whole again? What happened to sawn-off and discarded parts? Similar worries occurred in the case of transplant victims too, or those who had lost a leg in battle. Equally, there was a worry that if you had been subject to such practices, you might lose your identity for eternity.

The advances in transplant medicine were aided by the long wars between England and France in the late eighteenth and early nineteenth centuries. The Creature in *Frankenstein* is explicitly composed of the joined-together parts of animals and humans. He is a new being formed from the parts of others, who, by this means, would have been left incomplete. John Hunter famously experimented with transplant surgery, beginning, early in his career, by working out how to fit the healthy teeth of paupers into the rotten gums of rich people. In one bizarre experiment, he grafted a human tooth into a cockerel's comb to prove that tooth transplantation was possible. Teeth were a real issue at this time when dental care was not very sophisticated or well understood. Consequently, dentures were much in demand, but when made of human teeth they were incredibly expensive. As a direct result, thousands of men killed at the Battle of Waterloo had their teeth stolen by battleground scavengers. Dentures made from so-called 'Waterloo Teeth' were preferable to those that came from less healthy bodies. Eventually, Hunter made his name as an anatomist. When an elephant belonging to King George III's menagerie died, he was the one chosen to dissect it.

So few human bodies were available through legitimate means that the ones Hunter used in his anatomical research usually came from bodysnatchers. Resurrection men, as they were also known, would keep the teeth from the corpses and sell them separately. Bodysnatching was a huge problem in the eighteenth and early nineteenth centuries. People went to great lengths to protect their loved ones and their own corpses after death, including employing guards, making sure that nightwatchmen patrolled graveyards, using mortsafes, iron coffins and heavy stone slabs, all measures to make it more difficult to

FIG. 16 Caricature of body snatchers with an anatomist identified as William Hunter (*right*). W. Austin, etching with engraving, 1773.

steal bodies. If the body could be protected for long enough it would deteriorate until it was of no use to medical schools. Anatomists came under attack and there were riots at execution sites. In 1788 in New York a riot occurred after rumours that medical students were digging up the graves of paupers and black people. The mob's anger was directed at a medical doctor, Alexander Hamilton, and resulted in twenty deaths.

It was in this febrile atmosphere that William Burke and William Hare took the lucrative medical market for cadavers to its logical conclusion. Bodies were desperately needed, and the healthier the body the more money was paid for it. Burke and Hare decided to kill to provide Robert Knox, an anatomy lecturer at the University of Edinburgh, with the corpses he needed for demonstrations to medical students. Their killing spree started innocuously enough. When a lodger in Hare's house died on 29 November 1827 owing rent, the pair sold his body to Knox, receiving a handsome payment of £7 10s. Thereafter they did not wait until their victims had died from natural causes but suffocated them and handed over the body (*see Plate 30*). In total, they killed sixteen people over a ten-month period. In an ironic twist of fate, Hare turned King's evidence on the promise of immunity from prosecution, leaving Burke to be found guilty, hanged and dissected. Burke's skeleton can still be seen in the University of Edinburgh's Anatomical Museum.

The Burke and Hare case influenced the Anatomy Act of 1832, which John Abernethy helped bring about. This act allowed for the use of bodies that were unclaimed after forty-eight hours from hospitals, prisons and workhouses in anatomical dissections. Obviously the act affected poor people far more than the rich. While it effectively stopped the resurrectionist trade it did not stop the people's anger at anatomists. Riots continued into the 1830s. In Manchester, during the cholera epidemic, a man found that, following his grandson's death at Swan Street hospital, the head had been removed from his coffin. It transpired that a medical student had taken it, bribing the nurse to stay silent. A mob of thousands took to the streets, vandalizing the hospital. Trust in the medical profession suffered from these episodes. Perhaps anger at Frankenstein's medical objectivity – his lack of care for the ethics of his project – is one of the many lasting influences that Mary Shelley's novel has had on the public.

AFTERWORD

Many of the debates discussed in *Frankenstein* are complicated and continue to this day. A number of tricky questions have not yet been fully resolved, such as continuing issues of patient consent. We have a legacy of stories such as that of Henrietta Lack, a black woman who died of cervical cancer in the USA in 1951, whose cells were used without her, or her family's, knowledge to create the HeLa cell line used in medical research. Mary Shelley's novel questions whether scientific and medical progress justifies using animal and human subjects without their permission. The Creature censures Victor Frankenstein when he asks him, 'How dare you sport thus with life?'[1]

The lack of certainty over whether the Creature has a soul seems to anticipate our current worries about artificial intelligence. Will robots ever be able to replace humans? Will machines be able to do all that we can do, even creative and artistic tasks? People continue to worry that transplants and surgeries affect their sense of identity and self. 'Frankensteinian' continues to be used as shorthand to describe a particularly worrying kind of medicine. According to the US National Institution of Medicine, cloning is being driven by a so-called 'Frankenstein syndrome', meaning reckless experimentation on humans. The proposal to fuse together human cells with a rabbit egg was described as the creation of a 'Frankenbunny'. New technologies, such as CRISPR gene editing and stem-cell research cause similar anxieties. New forms of life are emerging all the time, and, as they do, the question of what rights they should have is raised.

We continue to build on our knowledge of how important 'vital' elements such as oxygen, water and climate are to life on earth. With the current climate crisis, people are asking, just as the Creature in *Frankenstein* does, whether access to fresh, clean water is a right that all humans should have. Equally, air quality is a live political issue today in the UK and globally. The British government now recognizes poor air quality as the largest environmental risk to public health in this country. It has led to an increase in chronic and sometimes fatal conditions such as asthma and lung cancer. We know that toxic particulate matter is breathed in by the population via fumes from vehicles, wood-burning stoves and other sources, particularly in urban areas. Mary Shelley already knew, what is patently obvious to us today, that clean air is vital to our well-being.

The framing narrative of the novel is set in the Arctic where Walton's ship is trapped in ice. In real life, in the summer of 1816, the long-range effects of a volcanic eruption led to terrible storms. In Europe, Mary Shelley witnessed refugees made homeless and starving because of the subsequent loss of crops. The questions that the novel asks about the effects of weather upon human mood and well-being continue to be asked today. For example, a team at the University of Manchester investigated oft-repeated claims that the weather exacerbates chronic pain.[2]

Drowning is the third leading cause of unintentional death worldwide and has been designated a major public-health problem. Many of the rudimentary resuscitation practices are still with us, from electric shocks to the rectal application of warm water (rather than tobacco!). But the success rate reported by the Royal Humane Society in 1818 seems fantastical by today's standards. According to the UK Resuscitation Council, English ambulance services initiate resuscitation on about 28,000 people each year, and of these some 8 per cent survive to leave hospital.[3] It is worth noting that Mary Shelley's novel performed a function in disseminating the most up-to-date methods of resuscitation at the time while it demonstrates to us her own medical knowledge.

Finally, and perhaps most disconcertingly, there is still discussion over how to define life and death. At what point a foetus is deemed to be alive continues to be a political issue for anti-abortion, 'pro-life' campaigners. But equally there are debates over whether death is the moment of 'brain death' or 'cardiac death'? The fact that the criteria and tests for confirming brain death are different in the USA and China reveal that these debates are far from over.

That current research was being represented in a fictional novel is fascinating. Mary Shelley's Frankenstein is a vitalist chemist who may (or may not) use electricity to bring a Creature to life. Personally, I do not think that Mary Shelley's novel demonstrates that she was against the idea of artificial creation per se. I think she is far more concerned with Victor's act of abandonment immediately afterwards. As a mother and a daughter she had already experienced such loss. If Frankenstein is punished in the novel, I think it is for this, rather than for usurping God's role as the creator of life. But my interpretation is, of course, coloured by my own historical moment and politics. Similarly, when I teach this novel, students are unanimously on the side of the Creature. They are convinced that he only acts in the way he does because of the way he is treated. *Frankenstein* teaches us that character is moulded by nurture not nature. It is our environment, our families and our friends that form us as human beings. While Mary

Shelley's parents, Mary Wollstonecraft and William Godwin, both emphasized the importance of environment on character, this is also a peculiarly modern sentiment. Finding out what men of science and medical practitioners thought at the time potentially challenges this view of the novel. Realizing that people thought they were bringing murderers back to life with electric shocks makes a different kind of sense of the fact that the Creature kills. This book reminds us of how people thought of 'nature' in the nature-versus-nurture debate. It reminds us of the reality of living and dying in the early nineteenth century.

NOTES

INTRODUCTION

1. The choice of how to name the protagonists in this book was a difficult one. While Percy Bysshe Shelley was known to his friends as 'Shelley' and Mary as 'Mary', this sits uncomfortably with modern readers who would bridle – understandably – at the idea that the 'Shelley' of this book is Percy Bysshe Shelley. Therefore, throughout, I have called them 'Mary' and 'Percy'.
2. Mary Shelley, *Frankenstein: 1818 Text*, ed. Nick Groom, Oxford University Press, Oxford, 2018, p. 176. I use this edition of *Frankenstein* throughout this book unless otherwise stated.
3. Mary Shelley, *The Journals of Mary Shelley: 1814–1844*, ed. Paula R. Feldman and Diana Scott-Kilvert, Johns Hopkins University Press, Baltimore MD and London, 1995, p. 55. Presumably this is the surgeon William Lawrence, who was embroiled in a very public debate in the Royal College of Surgeons with another surgeon, John Abernethy, on the nature of life (see Chapter 4).
4. Ibid., pp. 59, 60, 63, 61, 64.
5. Ibid., p. 65.
6. Ibid., pp. 65, 66.
7. Ibid., p. 68.
8. Mary Shelley, *The Letters of Mary Wollstonecraft Shelley*, ed. Betty T. Bennett, 2 vols, Johns Hopkins University Press, Baltimore MD, 1980, vol. 1, p. 11.
9. M. Shelley, *The Journals of Mary Shelley*, p. 68.
10. Ibid., p. 68. Mary misspells 'lose'.
11. M. Shelley, The *Letters of Mary Wollstonecraft Shelley*, vol. 1, p. 11.
12. M. Shelley, *The Journals of Mary Shelley*, pp. 74–8; p. 79.

13. Sharon Ruston, *Shelley and Vitality*, Palgrave Macmillan, Basingstoke, 2005, p. 91.
14. They read these aloud in French from *Fantasmagoriana, ou Recueil d'histoires d'apparitions de spectres, revenans, fantômes, etc.*, translated anonymously from the German by Jean-Baptiste Benoît Eyriès and published in 1812.
15. M. Shelley, *Frankenstein*, p. 175.
16. John Polidori, *The Diary of John William Polidori*, ed. William Michael Rossetti, Elkin Mathews, London, 1911, p. 123.
17. M. Shelley, *The Journals of Mary Shelley*, p. 56.
18. M. Shelley, *Frankenstein*, p. 34.
19. Mary Wollstonecraft, *The Collected Letters of Mary Wollstonecraft*, ed. Janet Todd, Allan Lane, London, 2003, pp. 327–8; p. 327.
20. Carolyn Williams first noted this pun, '"Inhumanly Brought Bach to Life and Misery": Mary Wollstonecraft, *Frankenstein*, and the Royal Humane Society', *Women's Writing*, vol. 8, no. 2, 2001, pp. 213–34; p. 232. Wollstonecraft, *The Collected Letters*, pp. 327–8; p. 327. Margaret Tims may have found further evidence of this suicide attempt in *The Times* report for 24 October 1795, which attributes the rescue of an unnamed 'Lady' to the 'skill of ... one of the medical persons belonging to the Humane Society'; *Mary Wollstonecraft: A Social Pioneer*, Millington, London, 1976, p. 273.
21. Wollstonecraft, *The Collected Letters*, pp. 326–7.
22. M. Shelley, *The Journals of Mary Shelley*, p. 70.
23. See, for example, James Curry, *Observations on Apparent Death from Drowning, Hanging, Suffocation by Noxious Vapours, Fainting-Fits, Intoxication, Lightning, Exposure to Cold, &c., &c. and an account of the proper means to be employed for recovery*, 2nd edn, E. Cox and Son, London, 1815, p. 78.
24. Percy Bysshe Shelley, *The Poems of Shelley*, ed. Geoffrey Matthews, Kelvin Everest, Jack Donovan, Cian Duffy and Michael Rossington, 4 vols, Longman, London and New York, 1989–2013, vol. 1, p. 401.
25. Percy Bysshe Shelley, *The Letters of Percy Bysshe Shelley*, ed. F.L. Jones, 2 vols, Clarendon Press, Oxford, 1964, vol. 2, p. 104, 25 July 1819.
26. M. Shelley, *Frankenstein*, pp. 40, 41, 49.
27. T. Cogan, *Memoirs of the Society Instituted at Amsterdam in favour of Drowned Persons, 1767–1771*, G. Robinson, London, 1773, p. iii.
28. Mary was not brought up in a very religious household. Her father, though raised as a Unitarian, was an atheist by the time of Mary's birth and agnostic later in his life.
29. Mary Wollstonecraft, *The Works of Mary Wollstonecraft*, ed. Janet Todd and Marilyn Butler, 7 vols, William Pickering, London, 1989, vol. 5, p. 193 n3.
30. Ibid., pp. 249, 218.
31. M. Shelley, *Frankenstein*, p. 173.

32. Anonymous, 'Review of *Frankenstein*', *Edinburgh Magazine, or Literary Miscellany* 2, 1818, pp. 249–53; p. 249.
33. Sharon Ruston, *Creating Romanticism: Case Studies in the Literature, Science and Medicine of the 1790s*, Palgrave Macmillan, Basingstoke, 2013, ch. 1.
34. Wollstonecraft, *The Works*, vol. 7, pp. 293–300; M. Shelley, *The Journals of Mary Shelley*, p. 39. Mary may have picked the book up on Claire Clairmont's recommendation because the month before the latter had spent a part of every day reading the *Philosophy of Natural History*. Claire Clairmont, *The Journals of Claire Clairmont*, ed. Marian Kingston Stocking, Harvard University Press, Cambridge MA, 1968, p. 46.
35. Wollstonecraft, *The Works*, vol. 7, p. 295; William Smellie, *The Philosophy of Natural History*, C. Elliot, T. Kay, T. Cadell and G.G.J. & J. Robinsons, Edinburgh and London, 1790, p. 44.
36. M. Shelley, *The Journals of Mary Shelley*, pp. 174, 117 n2. Percy Shelley ordered Buffon, *Le Théorie de la terre* (1749), volume 1 of the *Histoire Naturelle* (P.B. Shelley, *The Letters of Percy Bysshe Shelley*, vol. 1, p. 499).
37. P.B. Shelley, *The Letters of Percy Bysshe Shelley*, vol. 1, p. 499.
38. M. Shelley, *Frankenstein*, p. 25.
39. M. Shelley, *The Journals of Mary Shelley*, p. 121.
40. [William Godwin], *Report of Dr. Benjamin Franklin, and Other Commissioners, Charged by the King of France, with the Examination of Animal Magnetism, as now Practised at Paris. Translated from the French. With an Historical Introduction*, J. Johnson, London, 1785; Ruston, *Creating Romanticism*, ch. 2.
41. William Godwin, *Lives of the Necromancers*, Frederick J. Mason, London, 1834, p. v.
42. M. Shelley, *Frankenstein*, p. 29.
43. Ibid., p. 23.
44. Ibid., p. 24; Mary Shelley, *Frankenstein*, ed. M.K. Joseph, Oxford University Press, Oxford, 1968, repr. 2008, p. 41.
45. M. Shelley, *Frankenstein*, pp. 23, 29.
46. P.B. Shelley, *The Letters of Percy Bysshe Shelley*, vol. 1, p. 319; Percy Bysshe Shelley, *The Witch of Atlas Notebook: A Facsimile of Bodleian MS. Shelley adds. e. 6*, ed. Carlene A. Adamson, Garland Publishing, New York, 1997.
47. M. Shelley, *The Journals of Mary Shelley*, pp. 142–4, 143.
48. Laura E. Crouch, 'Davy's *A Discourse, Introductory to A Course of Lectures on Chemistry*: A Possible Scientific Source of *Frankenstein*', *Keats–Shelley Journal* 27, 1978, pp. 35–44.
49. M. Shelley, *Frankenstein*, p. 29.
50. Humphry Davy, *The Collected Works of Humphry Davy*, ed. John Davy, 9 vols, Smith, Elder and Co., London, 1839, vol. 2, p. 319.
51. Charles E. Robinson, *The Frankenstein Notebooks: A Facsimile Edition of*

Mary Shelley's Manuscript Novel, 1816–17, 2 vols, Garland, New York, 1996, vol. 1, p. lxvii. The surviving *Frankenstein* manuscripts (and others by the Shelleys' circle) have been digitized and are freely available online at the Godwin Shelley Archive: shelleygodwinarchive.org. Readers might also be interested in the digitized William Godwin: Diarygodwindiary.bodleian.ox.ac.uk/index2.html?.
52. Oxford, Bodleian Library, MS. Abinger c. 56, fol. 4r.
53. M. Shelley, *Frankenstein*, p. 21.
54. Oxford, Bodleian Library, MS. Abinger c. 56, fols 1r, 5v, 3v, 11v.
55. William Whewell proposed the term 'scientist' 'by analogy with *artist*': [William Whewell], '*On the Connexion of the Physical Sciences*. By Mrs. Somerville', *Quarterly Review* 51, 1834, pp. 54–68; p. 59.
56. M. Shelley, *Frankenstein*, pp. 30, 31.
57. Ibid., pp. 45, 116, 120, 129.
58. He had also played Ruthven in a stage version of Polidori's *The Vampyre* a few years before in 1820. The connections between Polidori's and Mary Shelley's texts are discussed in Chapter 1.
59. Richard Brinsley Peake, *Presumption; or, the Fate of Frankenstein (1823)*, Act I, digital edn, ed. Stephen C. Behrendt, 2001, https://romantic-circles.org/editions/peake/index.html.
60. M. Shelley, *Frankenstein*, p. 35.

ONE
1. M. Shelley, *Frankenstein*, p. 176.
2. Anna Letitia Barbauld, 'A Mouse's Petition', in *The Poems of Anna Letitia Barbauld*, ed. William McCarthy and Elizabeth Kraft, University of Georgia Press, Athens GA, 1994, pp. 36–7, line 25.
3. M. Shelley, *Frankenstein*, p. 70.
4. Barbauld, 'A Mouse's Petition', lines 21, 24.
5. Samuel Taylor Coleridge, *Samuel Taylor Coleridge: The Major Works*, ed. H.J. Jackson, Oxford University Press, Oxford and New York, 1985, p. 28, line 27.
6. Barbauld, 'A Mouse's Petition', line 28.
7. Barbauld, 'Life', in *The Poems*, p. 166, lines 7, 8.
8. M. Shelley, *The Journals of Mary Shelley*, pp. 25, 31.
9. M. Shelley, *Frankenstein*, p. 12, 38, 114.
10. Coleridge, The Rime of the Ancient Mariner, in *Coleridge: The Major Works*, pp. 47, 52–3, 57, 64, lines 11, 193–8, 214, 339–40, 584.
11. *Frankenstein*, dir. Danny Boyle, National Theatre, 2018, http://ntlive.nationaltheatre.org.uk/productions/ntlin4-frankenstein.
12. Letter from Humphry Davy to Samuel Taylor Coleridge, 26 November 1800, in *The Collected Letters of Sir Humphry Davy*, ed. Tim Fulford and

Sharon Ruston, 4 vols, Oxford University Press, Oxford, 2020, vol. 1, letter 33.
13. Samuel Taylor Coleridge, *Biographia Literaria*, 2 vols, ed. J. Engell and W. Jackson Bate, Princeton University Press, Princeton NJ, 1993, vol. 1, p. 304.
14. Ibid., vol. 2, p. 4.
15. Letter to Sir Humphry Davy, 15 July 1800, in *The Collected Letters of Samuel Taylor Coleridge*, ed. E.L. Griggs, 5 vols, Clarendon Press, Oxford, 1956–71, vol. 1, p. 605.
16. Samuel Taylor Coleridge, *Hints Towards the Formation of a More Comprehensive Theory of Life, 1818*, ed. S.B. Watson, John Churchill, London, 1848.
17. John Abernethy, The *Hunterian Oration, for the year 1819*, Longman, Hurst, Rees, Orme, & Browne, London, 1819, p. 66.
18. Coleridge, *Hints*, pp. 38, 42.
19. Coleridge, *Biographia Literaria*, p. 16.
20. Coleridge, 'Human Life, On the Denial of Immortality, A Fragment', in *The Complete Poetical Works*, ed. Ernest Hartley Coleridge, 2 vols, Clarendon Press, Oxford, 1912, repr. 1975, vol. 1, p. 425, lines 4–5.
21. Ibid., lines 5–6, 3, 8–9, 29.
22. Lord Byron, *Lord Byron: The Complete Poetical Works*, Volume 5: *Don Juan*, ed. Jerome J. McGann, 7 vols, Clarendon Press, Oxford, 1986, p. 50, canto 1, cxxx, lines 1034–7.
23. M. Shelley, *Frankenstein*, p. 175.
24. Lord Byron, *Manfred*, in *The Complete Poetical Works*, vol. 4, 73, 2.2.79–83.
25. Ibid., p. 68, 2.1.38; 2.2.57.
26. Ibid., p. 72, 1.2.40, 41.
27. Prometheus features in William Godwin's pseudonymously published *The Pantheon: or Ancient History of the Gods of Greece and Rome*, T. Hodgkins, London, 1806.
28. Lord Byron, 'Prometheus', in *The Complete Poetical Works*, vol. 4, p. 32, line 45.
29. Ibid., line 35.
30. M. Shelley, *Frankenstein*, p. 34.
31. Ibid., p. 33.
32. For evidence of Byron and Polidori's sexual relationship, see for example Peter Cochran, 'Byron's Boyfriends', in Peter Cochran (ed.), *Byron and Women [and Men]*, Cambridge Scholars Press, Newcastle upon Tyne, 2010, pp. 15–56; p. 52.
33. Polidori, *The Diary of John William Polidori*, p. 123.
34. D.L. Macdonald, *Poor Polidori: A Critical Biography of the Author of 'The Vampyre'*, University of Toronto Press, Toronto, 1991, p. 20.

35. M. Shelley, *Frankenstein*, p. 52.
36. Lord Byron, 'The Giaour', in *The Complete Poetical Works*, vol. 3, p. 64, lines 760, 762.
37. Polidori, *The Diary of John William Polidori*, p. 128.
38. John William Polidori, *The Vampyre and Ernestus Berchtold; or, the Modern Oedipus*, ed. D.L. Macdonald and Kathleen Scherf, Broadview Press, Peterborough ON, p. 39.
39. Ibid., p. 41.
40. Polidori, *The Diary of John William Polidori*, p. 119.
41. Erasmus Darwin, The *Temple of Nature; or, the Origin of Society, a Poem, with Philosophical Notes*, John W. Butler, Baltimore MD, 1804, p. 139.
42. Macdonald, *Poor Polidori*, p. 77.
43. S.T. Coleridge, *Christabel &c.*, 3rd edn, John Murray, London, 1816, p. 52.
44. M. Shelley, *Frankenstein*, pp. 175–6.
45. P.B. Shelley, *The Letters of Percy Bysshe Shelley*, vol. 1, p. 121, July 1811.
46. Ruston, *Shelley and Vitality*, ch. 2.
47. Percy Bysshe Shelley, 'On Life', in *Shelley's Poetry and Prose*, ed. Donald H. Reiman, W.W. Norton, New York, 1977, p. 505.
48. P.B. Shelley, 'Mutability', in *The Poems of Shelley*, vol. 1, p. 456, line 2.
49. Ibid., lines 13, 16.
50. P.B. Shelley, Alastor; Or, the Spirit of Solitude *Poems*, in *The Poems of Shelley*, vol. 1, p. 465, lines 23–4, 26, 29.
51. P.B. Shelley, 'Hymn to Intellectual Beauty', in *The Poems of Shelley*, vol. 1, p. 526, line 30.
52. P.B. Shelley, 'Mont Blanc', in *The Poems of Shelley*, vol. 1, p. 540, lines 95–6.
53. Percy Bysshe Shelley, *A Vindication of Natural Diet*, F. Pitman, London, 1884, p. 26.
54. P.B. Shelley, Adonais, in *The Poems of Shelley*, vol. 4, p. 314, lines 348–9.
55. Mary Shelley, *The Mary Shelley Reader*, ed. Betty T. Bennett and Charles E. Robinson, Oxford University Press, Oxford, 1990, p. 336.
56. P.B. Shelley, Adonais, p. 315, line 370.
57. Thomas Charles Morgan, *Sketches of the Philosophy of Life*, Henry Colburn, London, 1818, p. 50.
58. John Keats, 'Ode to a Nightingale', in *John Keats: The Major Works*, ed. Elizabeth Cook, Oxford University Press, Oxford, 2008, pp. 285–8, lines 79, 80, 1, 21, 26.
59. John Keats, 'This Living Hand', in *John Keats: The Major Works*, p. 331, lines 1, 4, 7–8.

TWO

1. M. Shelley, *Frankenstein*, p. 14.
2. Ibid., p. 42, 47.
3. Ibid., pp. 91, 97, 100, 148.
4. Ibid., p. 113.
5. Ibid., p. 24.
6. Joseph Priestley, *Experiments and Observations on Different Kinds of Air*, 6 vols, J. Johnson, London, 1774–86, vol. 1, p. xiv.
7. M. Shelley, *Frankenstein*, p. 29.
8. Ibid., p. 30.
9. Ibid., p. 74.
10. Ibid., p. 102.
11. Ibid., p. 175.
12. Shelley, *The Witch of Atlas Notebook*, vol. 5, pp. 167, 166 *rev.*; Humphry Davy, *Elements of Agricultural Chemistry*, 2nd edn, Longman, Hurst, Rees, Orme, & Browne, London, 1814, p. 44.
13. P.B. Shelley, *The Witch of Atlas Notebook*, vol. 5, p. 164 *rev.*
14. Ibid., pp. 171–70 *rev.*
15. M. Shelley, *Frankenstein*, p. 103. The Creature speaks of how the bright day 'cheered even me by the loveliness of its sunshine'; ibid.
16. Ibid., p. 108.
17. Anonymous, *Annual Report of the Royal Humane Society for the Recovery of Persons Apparently Drowned or Dead*, Printed for the Society, London, 1818, p. 56.
18. Ibid., p. 57.
19. M. Shelley, *Frankenstein*, pp. 136, 135.
20. *Annual Report of the Royal Humane Society*, p. 63.
21. M. Shelley, *Frankenstein*, pp. 156, 139, 147, 8–9.
22. Ibid., p. 64.
23. Ibid., p. 130.
24. Ibid., p. 104.
25. Ibid., pp. 140, 133.
26. Hunter believed that the vital principle was to be found in the blood. John Hunter, 'Proposals for the Recovery of People Apparently Drowned', *Philosophical Transactions* 66, 1776, pp. 412–25; p. 414n.). Percy Shelley ordered a book that contained Hunter's essay, Robert Thornton's *The Philosophy of Medicine: being Medical Extracts on the Nature and Preservation of Health, and on the Nature and Removal of Disease* (2 vols, 5th edn, Sherwood, Neely and Jones, London, 1813, vol. 2, pp. 52–5), on 29 July 1812; P.B. Shelley, *The Letters of Percy Bysshe Shelley*, vol. 1, p. 319.
27. Godwin's diary for 5 November 1817 reads: 'Dr Curry calls (on S.)', and for 13 November 1817: 'dine at Shelley's; adv. Curry, Baxter & Booth'; *William*

Godwin's Diary, Oxford, Bodleian Library, MS. Abinger e. 20, fol. 37v. M. Shelley, *The Journals of Mary Shelley*, vol. 1, p. 47, 24 November 1814.
28. Curry, *Observations on Apparent Death*, ch. 1.
29. Ibid., pp. ivn, vn.
30. Ibid., pp. 1, 4.
31. Ibid., pp. 15, 13.
32. Ibid., p. 206.
33. Wollstonecraft, The *Collected Letters*, pp. 327–8; p. 327.
34. Hanging killed by asphyxiating rather than by dislocating the spine before the introduction of the 'new drop' gallows in London's Newgate Prison in 1783, which had a trapdoor that opened below the body.
35. Laura Gowing, 'Greene, Anne (*c.* 1628–1659)', *ODNB*.
36. Phillipe Ariès, *The Hour of Our Death*, trans. Helen Weaver, Allen Lane, London, 1981, p. 400.
37. John Davy, *Memoirs of the Life of Sir Humphry Davy*, 2 vols, Longman, Rees, Orme, Brown, and Green, London, 1836, vol. 2, p. 368.
38. M. Shelley, *Frankenstein*, p. 162.
39. Ibid., p. 139.
40. Ibid., pp. 136, 48, 133, 149.
41. Curry, *Observations on Apparent Death*, p. 70.
42. M. Shelley, *Frankenstein*, pp. 34, 35.
43. Ibid., pp. 110–11.
44. Curry, *Observations on Apparent Death*, p. 130.
45. P.B. Shelley, *Letters of Percy Bysshe Shelley*, vol. 2, p. 78.
46. M. Shelley, *Frankenstein*, p. 108.
47. Ibid., p. 53.
48. Ibid., p. 171.

THREE
1. M. Shelley, *Frankenstein*, p. 24.
2. Ibid., p. 33.
3. Ibid., p. 175.
4. Ibid., pp. 36, 70, 99.
5. Benjamin Franklin, 'A letter from Mr. Franklin to Mr. Peter Collinson, F.R.S. concerning the effects of lightning', *Philosophical Transactions* 47, 1752, pp. 289–91.
6. Benjamin Franklin, 'A letter of Benjamin Franklin, Esq; to Mr. Peter Collinson, F.R.S. concerning an electrical kite', *Philosophical Transactions* 47, 1752, pp. 565–7.
7. Thomas Jefferson Hogg, *The Life of Percy Bysshe Shelley*, 2 vols, Edward Moxon, London, 1858, vol. 1, p. 70. It should be noted that Hogg's memoirs

were notoriously unreliable, but he is one of only a few people who knew Percy at university.
8. P.B. Shelley, *The Witch of Atlas Notebook*, p. 166 *rev*. 'Electy' is an abbreviation for 'Electricity'.
9. Adam Walker, *A System of Familiar Philosophy: In Twelve Lectures; Being the Course of Lectures Usually Read by Mr. A. Walker*, 2 vols, 2nd edn, G. Kearsley, London, 1802, vol. 1, p. 142.
10. Thomas Medwin, *The Life of Percy Bysshe Shelley*, 2 vols, Thomas Cautley Newby, London, 1847, vol. 1, p. 112.
11. M. Shelley, *Frankenstein*, p. 24.
12. Oxford, Bodleian Library, MS. Abinger c. 56, fol. 5v.
13. Oxford, Bodleian Library, MS. Abinger c. 56, fol. 5r, fol. 3r, fol. 2v.
14. M. Shelley, *Frankenstein*, p. 25.
15. Oxford, Bodleian Library, MS. Abinger c. 56, fol. 3v.
16. M. Shelley, *Frankenstein*, p. 25.
17. Oxford, Bodleian Library, MS. Abinger c. 56, fol. 3v, fol. 11v.
18. Hogg, *The Life of Percy Bysshe Shelley*, vol. 1, pp. 8–9.
19. Ibid., pp. 33, 41, 62.
20. Coincidently, Humphry Davy died in Geneva and was buried in Plainpalais in 1829.
21. M. Shelley, *Frankenstein*, p. 51.
22. Ibid., p. 52.
23. This scene is one that can be traced to a letter Mary wrote to Fanny Imlay; Mary Shelley and P.B. Shelley, *History of A Six Weeks' Tour*, Woodstock Books, Oxford, 1989, pp. 99–100.
24. M. Shelley, *Frankenstein*, p. 8.
25. John Aldini, *An Account of the Late Improvements in Galvanism…*, J. Murray, London, 1803, pp. 61, 67, 68.
26. M. Shelley, *Frankenstein*, p. 35.
27. Ibid., p. 32; Aldini, *An Account*, p. 68.
28. M. Shelley, *Frankenstein*, p. 32.
29. Aldini, *An Account*, p. 68.
30. M. Shelley, *Frankenstein*, pp. 113, 124.
31. Ibid., pp. 37, 38.
32. Ibid., p. 38.
33. Ibid., p. 95.
34. Anonymous, 'Art. XV. *An Account of the late Improvements in Galvanism…*', *Edinburgh Review* 3, 1803, pp. 194–8; p. 196.
35. M. Shelley, *Frankenstein*, p. 173.
36. Ibid., p. 36.
37. Aldini, *An Account*, p. 193.
38. Ibid., p. 54.

39. Ibid., p. 68.
40. Andrew Knapp, 'George Forster', *The New Newgate Calendar...*, 5 vols, J. Robins & Co., London, 1826, vol. 4, pp. 182–9; p. 189.
41. M. Shelley, *Frankenstein*, p. 35.
42. Andrew Ure, 'An Account of Some Experiments made on the Body of a Criminal immediately after Execution, with Physiological and Practical Observations', *Quarterly Journal of Science 6*, 1819, pp. 283–94; p. 290.
43. Aldini, *An Account*, pp. 110–12.
44. Ibid., p. 95.
45. Ibid., p. 91.
46. Ibid., p. 53.
47. M. Shelley, *Frankenstein*, pp. 29–30.
48. Charles Kite, *Essay in the Recovery of the Apparently Dead*, C. Dilly, London, 1788, p. 123.
49. Ibid., p. 123.
50. Ibid., p. 126.
51. Aldini, *An Account*, p. 130.
52. Adam Walker, *Analysis of a Course of Lectures on Natural and Experimental Philosophy*, 14th edn, J. Barfield, London, 1807, pp. 55, 57–8.
53. Hogg, *The Life of Percy Bysshe Shelley*, vol. 1, p. 9; Walker, *A System*, vol. 2, p. 50.
54. John Wesley, *The Desideratum: Or, Electricity made Plain and Useful by a Lover of Mankind and of Common Sense*, W. Flexney, London, 1760.
55. Erasmus Darwin, *The Botanical Garden*, *1791*, 2 vols, Scholar Press, Menston, 1973, vol. 1, p. 435.
56. Walker, *A System*, vol. 2, p. 73.
57. [Godwin], *Report*. See Ruston, *Creating Romanticism*, ch. 2.
58. [Godwin], *Report*, p. xv.
59. Mary Wollstonecraft, *A Vindication of the Rights of Woman*, in *Works*, vol. 5, pp. 253, 185n.
60. John Cordy Jeaffreson, *The Real Shelley*, 2 vols, Hurst and Blackett, London, 1885, vol. 1, p. 21. Jeaffreson's biography should be treated with caution since it is not always reliable.
61. Ibid., pp. 21–2.
62. George Colman, *The Plays of George Colman The Elder*, ed. Kalman A. Burnim, 6 vols, Garland Publishing, New York and London, 1983, vol. 5, p. 11.
63. Ibid.
64. James Graham M.D., *The General State of Medical and Chirurgical Practice, exhibited, shewing them to be Inadequate, Ineffectual, Absurd and Ridiculous*, 6th edn, Mr. Almond and Messrs Richardson & Urquhart, London, 1779, p. 53.

65. Letter from Humphry Davy to Samuel Taylor Coleridge, 26 November 1800, in *The Collected Letters of Sir Humphry Davy*, vol. 1, letter 33.
66. Davy, *Collected Works*, vol. 2, p. 86.

FOUR

1. M. Shelley, *Frankenstein*, pp. 32, 33.
2. Anonymous, *The Times*, 'Law Report', *Lawrence v. Smith*, 25 March 1822, p. 3.
3. M. Shelley, *Frankenstein*, p. 175.
4. Mary Shelley might have meant 'the vorticella, or wheeled animal', rather than 'vermicelli', mentioned in Erasmus Darwin, *The Temple of Nature*, Additional Notes, p. 7.
5. Darwin, *The Temple of Nature*, p. 21.
6. M. Shelley, *Frankenstein*, p. 175.
7. G.L.L. Buffon, *Natural History, General and Particular, by the Count de Buffon, Translated into English*, 8 vols, William Creech, Edinburgh, 1780, vol. 1, p. 1.
8. J.F. Blumenbach, *An Essay on Generation*, trans. A. Crichton, T. Cadell, London, 1792, p. 20.
9. Ibid.
10. Samuel Taylor Coleridge, *The Notebooks of Samuel Taylor Coleridge*, ed. K. Coburn, 5 vols, Routledge & Kegan Paul, London, 2002, vol. 3, p. 3744.
11. P.B. Shelley, The *Letters of Percy Bysshe Shelley*, vol. 2, p. 361.
12. Medwin, *The Life of Percy Bysshe Shelley*, vol. 1, p. 36.
13. P.B. Shelley, *Shelley's Prose, or the Trumpet of a Prophecy*, ed. David Lee Clark, University of New Mexico Press, Albuquerque NM, 1966, p. 274.
14. M.Shelley, *Frankenstein*, p. 175.
15. John Abernethy, *An Enquiry into the Probability and Rationality of Mr. Hunter's Theory of Life: being the subject of the first two anatomical lectures delivered at the Royal College of Surgeons, of London*, Longman, Hurst, Rees, Orme, & Brown, London, 1814, p. 1.
16. Ibid., p. 3.
17. John L. Thornton, *John Abernethy: A Biography*, Printed by the author, London, 1953, p. 58.
18. The note written by Shelley reads 'see Abernethy'; Percy Bysshe Shelley, *Miscellaneous Poetry, Prose and Translations from Bodleian MS. Shelley adds. c.4, Folios 63, 65, 71, and 72*, ed. E.B. Murray, Garland Publishing, New York and London, 1992, adds. c.4, fol. 272r.
19. John Abernethy, *Surgical Observation on the Constitutional Origin and Treatment of Local Diseases; and on Aneurisms*, Longman, Hurst, Rees, Orme, & Brown, London, 1807.
20. Hogg, The *Life of Percy Bysshe Shelley*, vol. 2, pp. 552–3.

21. *William Godwin's Diary*, 14 January 1813, Oxford, Bodleian Library, MS. Abinger e. 17, fol. 23r.
22. J.L. Thornton, 'John Abernethy 1764–1831', *St Bartholomew's Hospital Journal* 72, July 1964, pp. 287–93; p. 289. Byron might be referring to this aspect of Abernethy's reputation when he describes him as 'soft Abernethy' in canto X of *Don Juan*; Lord Byron, *Don Juan, The Complete Poetical Works*, vol. 5, p. 449, canto X, stanza 42, line 336.
23. George Macilwain, *Memoirs of John Abernethy*, 3rd edn, Hatchard, London, 1856, p. 190.
24. Letter to J.H. Green, 25 May 1820, Coleridge, *Collected Letters of Samuel Taylor Coleridge*, vol. 5, p. 49 n2.
25. Anonymous, 'Sir William Lawrence, Bart', *St. Bartholomew's Hospital Reports* 4, 1868, pp. 1–18; p. 2.
26. June Goodfield-Toulmin, 'Some Aspects of English Physiology: 1780–1840', *Journal of the History of Biology* 2, 1969, pp. 283–320; p. 319.
27. Letter to Percy Shelley, 26 September 1817, in *The Letters of Mary Wollstonecraft Shelley*, vol. 1, p. 43.
28. *William Godwin's Diary*, 1 June 1812, 11 November 1812, 5 March 1813, 18 March 1815, 4 March 1818.
29. Abernethy, *An Enquiry*, p. 5.
30. M. Shelley, *Frankenstein*, p. 32.
31. Abernethy, *An Enquiry*, p. 16.
32. John Hunter, *A Treatise on the Blood, Inflammation, and Gun-shot Wounds...*, George Nicol, London, 1794, p. 78.
33. Leviticus 17:2.
34. M. Shelley, *Frankenstein*, pp. 63, 149.
35. Abernethy, *An Enquiry*, p. 17.
36. Ibid., p. 28.
37. Ibid., p. 32.
38. Ibid., pp. 39, 41.
39. Ibid., pp. 48, 49.
40. Ibid., pp. 49, 51.
41. M. Shelley, *Frankenstein*, p. 32.
42. Ibid., p. 24.
43. Ibid., p. 102.
44. Ibid., p. 37.
45. Ibid., p. 33.
46. Ibid.
47. Abernethy, *An Enquiry*, pp. 87–8.
48. Ibid., p. 95.
49. Xavier Bichat, *Physiological Researches on Life and Death*, trans. F. Gold, Longman, Hurst, Rees, Orme, & Brown, London, 1816, p. 21.

50. Ibid., p. 22.
51. Ibid., p. 23.
52. William Lawrence, *An Introduction to Comparative Anatomy and Physiology…*, J. Callow, London, 1816, p. 2.
53. Ibid., p. 3.
54. Ibid., p. 15.
55. M. Shelley, *Frankenstein*, p. 34.
56. Lawrence, *An Introduction*, p. 87. To rectify this, Humphry Davy instigated the London Zoo, planned as a direct counterpart to the Jardin des Plantes in 1826.
57. Ibid., p. 88.
58. M. Shelley, *Frankenstein*, p. 5.
59. Ibid., p. 113.
60. Ibid., p. 119.
61. Lawrence, *An Introduction*, p. 112.
62. M.Shelley, *Frankenstein*, p. 162.
63. Lawrence, *An Introduction*, pp. 117, 121.
64. M. Shelley, *Frankenstein*, pp. 32, 48.
65. Lawrence, *An Introduction*, p. 128.
66. Ibid., p. 129.
67. Ibid., pp. 130–31.
68. Ibid., p. 139.
69. Ibid., pp. 163, 169.
70. Ibid., p. 170.
71. Ibid., p. 174.
72. Ibid., p. 179.
73. John Abernethy, *Physiological Lectures, Exhibiting a General View of Mr Hunter's Physiology, and of his researches in Comparative Anatomy, delivered before the Royal College of Surgeons, in the year 1817*, Longman, Hurst, Rees, Orme & Brown, London, 1817, p. 27.
74. Ibid., pp. 36, 37, 38.
75. Ibid., p. 46.
76. Ibid., p. 47; William Shakespeare, *Macbeth*, 3.4.78 (*The Oxford Shakespeare: The Complete Works*, 2nd edn, Oxford University Press, Oxford, 2005, p. 982).
77. Abernethy, *Physiological Lectures*, pp. 49, 50, 52.
78. William Lawrence, *Lectures on Physiology, Zoology, and the Natural History of Man*, W. Callow, London, 1819.
79. Thomas Rennell, *Remarks on Scepticism, Especially as it is Connected with the Subjects of Organization and Life. Being an Answer to the Views of M. Bichat, Sir T. C. Morgan and Mr. Lawrence upon these Points*, F.C. and J. Rivington, London, 1819, p. 64.

80. Edward Grinfield, *Cursory Observations on the 'Lectures on Physiology, Zoology, and the Natural History of Man, delivered at the Royal College of Surgeons by William Lawrence F. R. S. Professor of Anatomy and Surgery to that College, &c. &c. &c.' In a series of letters addressed to that Gentleman; with a Concluding Letter to his Pupils*, 2nd edn, T. Cadell and W. Davies, London, 1819, p. 4.
81. [George D'Oyly], 'An Enquiry into the Probability of Mr. Hunter's Theory of Life...', *Quarterly Review* 43, 1819, pp. 1–34; p. 2.
82. Lawrence, *Lectures on Physiology*, p. 7.
83. *The Times*, 25 March 1822, p. 3.
84. Anonymous, *Monthly Magazine* 53, 1822, pp. 524–44; p. 542.
85. M. Shelley, *Frankenstein*, p. 91.

FIVE

1. M. Shelley, *Frankenstein*, p. 37.
2. Ibid., p. 32.
3. Ibid., pp. 36, 134, 141, 149, 168, 169.
4. Ibid., p. 32.
5. Ibid., p. 32.
6. Anonymous, 'An Inquiry into the Probability and Rationality of Mr. Hunter's Theory of Life; being the subject of the first Two Anatomical Lectures delivered before the Royal College of Surgeons, of London. By John Abernethy', *Edinburgh Review* 23, 1814, pp. 384–98; pp. 384–5.
7. M. Shelley, *Frankenstein*, p. 34.
8. Ibid., p. 34.
9. Ibid., p. 32.
10. Humphry Davy, *Elements of Chemical Philosophy*, in *The Collected Works of Humphry Davy*, vol. 4, p. 127.
11. William Brande, 'Observations on Albumen, and Some Other Animal Fluids; with Remarks on their Analysis by Electro-Chemical Decomposition', *Philosophical Transactions* 99, 1809, pp. 373–84; Everard Home, 'Hints on the Subject of Animal Secretions', *Philosophical Transactions* 99, 1809, pp. 385–91; William Brande, 'Chemical Researches on the Blood, and some other Animal Fluids', *Philosophical Transactions* 102, 1812, pp. 90–104; William Brande, 'Observations on the Effects of Magnesia, in Preventing an Increased Formation of Uric Acid; with Some Remarks on the Composition of the Urine', *Philosophical Transactions* 100, 1810, pp. 136–47; B.C. Brodie, 'Experiments and Observations on the influence of the Nerves of the eighth pair on the secretions of the stomach', *Philosophical Transactions* 104, 1814, pp. 102–6; B.C. Brodie, 'Experiments and Observations on the Different Modes in which Death is Produced by Certain Vegetable Poisons', *Philosophical Transactions* 101, 1811, pp. 178–208; B.C. Brodie,

'Further Experiments and Observations on the Influence of the Brain on the Generation of Animal Heat', *Philosophical Transactions* 102, 1812, pp. 378–93.
12. P.B. Shelley, *The Witch of Atlas Notebook*, pp. 171–70 *rev*.
13. M. Shelley, *Frankenstein*, p. 23.
14. Ibid., p. 151.
15. John Brown, *Elements of Medicine: A New Edition*, trans. Thomas Beddoes, 2 vols, J. Johnson, London, 1795, vol. 1, p. cxxvii.
16. Ibid., p. cxxvii.
17. William Wordsworth, The *Prose Works of William Wordsworth*, ed. W.J.B. Owen and J.W. Smyser, 3 vols, Clarendon Press, Oxford, 1974, vol. 1, p. 150.
18. Coleridge, *Biographia Literaria*, vol. 2, p. 229.
19. M. Shelley, *Frankenstein*, p. 140.
20. Ruston, *Creating Romanticism*, pp. 117–20.
21. M. Shelley, *Frankenstein*, p. 32.
22. Ibid., p. 36.
23. Ibid., p. 35.
24. Ibid., pp. 40, 134, 40.
25. Ibid., p. 44.
26. Ibid., p. 45.
27. Ibid., pp. 163, 166, 167.
28. Ibid., p. 171.

AFTERWORD
1. M. Shelley, *Frankenstein*, p. 70.
2. Cloudy with a Chance of Pain, www.cloudywithachanceofpain.com.
3. Resuscitation Council UK, www.resus.org.uk/resuscitation-guidelines/introduction.

FURTHER READING

Abernethy, J., *Physiological Lectures, Exhibiting a General View of Mr Hunter's Physiology, and of his researches in Comparative Anatomy, delivered before the Royal College of Surgeons, in the year 1817*, Longman, Hurst, Rees, Orme, & Brown, London, 1817.

Abernethy, J., *An Enquiry into the Probability and Rationality of Mr. Hunter's Theory of Life: being the subject of the first two anatomical lectures delivered at the Royal College of Surgeons, of London*, Longman, Hurst, Rees, Orme, & Brown, London, 1814.

Abernethy, J., *Surgical Observation on the Constitutional Origin and Treatment of Local Diseases; and on Aneurisms*, Longman, Hurst, Rees, Orme, & Brown, London, 1807.

Aldini, J., *An Account of the Late Improvements in Galvanism…*, J. Murray, London, 1803.

Anonymous, 'Sir William Lawrence, Bart', *St. Bartholomew's Hospital Reports* 4, 1868, pp. 1–18.

Anonymous, *Monthly Magazine* 53, 1822, pp. 524–44.

Anonymous, *The Times*, 'Law Report', Lawrence *v.* Smith, 25 March 1822, p. 3.

Anonymous, *Annual Report of the Royal Humane Society for the Recovery of Persons Apparently Drowned or Dead*, Printed for the Society, London, 1818.

Anonymous, 'Review of *Frankenstein*', *Edinburgh Magazine, or Literary Miscellany* 2, 1818, pp. 249–53.

Anonymous, 'An Inquiry into the Probability and Rationality of Mr. Hunter's Theory of Life; being the subject of the first Two Anatomical Lectures delivered before the Royal College of Surgeons, of London. By John Abernethy', *Edinburgh Review* 23, 1814, pp. 384–98.

Anonymous, 'Art. XV. *An Account of the late Improvements in Galvanism…*', *Edinburgh Review* 3, 1803, pp. 194–8.

Bichat, X., *Physiological Researches on Life and Death*, trans. F. Gold, Longman, Hurst, Rees, Orme, & Brown, London, 1816.

Blumenbach, J. F., *An Essay on Generation*, trans. A. Crichton, T. Cadell, London, 1792.
Brande, W., 'Chemical Researches on the Blood, and some other Animal Fluids', *Philosophical Transactions* 102, 1812, pp. 90–104.
Brande, W., 'Observations on the Effects of Magnesia, in Preventing an Increased Formation of Uric Acid; with Some Remarks on the Composition of the Urine', *Philosophical Transactions* 100, 1810, pp. 136–47.
Brande, W., 'Observations on Albumen, and Some Other Animal Fluids; with Remarks on their Analysis by Electro-Chemical Decomposition', *Philosophical Transactions* 99, 1809, pp. 373–84.
Brodie, B.C., 'Experiments and Observations on the influence of the Nerves of the eighth pair on the secretions of the stomach', *Philosophical Transactions* 104, 1814, pp. 102–6.
Brodie, B.C., 'Further Experiments and Observations on the Influence of the Brain on the Generation of Animal Heat', *Philosophical Transactions* 102, 1812, pp. 378–93.
Brodie, B.C., 'Experiments and Observations on the Different Modes in which Death is Produced by Certain Vegetable Poisons', *Philosophical Transactions* 101, 1811, pp. 178–208.
Brown, J., *Elements of Medicine: A New Edition*, trans. Thomas Beddoes, 2 vols, J. Johnson, London, 1795.
Buffon, G.L.L., *Natural History, General and Particular, by the Count de Buffon, Translated into English*, 8 vols, William Creech, Edinburgh, 1780.
Butler, M., 'The First Frankenstein and Radical Science', *Times Literary Supplement*, 9 April 1993, pp. 12–14.
Clairmont, C., *The Journals of Claire Clairmont*, ed. Marian Kingston Stocking, Harvard University Press, Cambridge MA, 1968.
Cloudy with a Chance of Pain, www.cloudywithachanceofpain.com.
Cogan, T., *Memoirs of the Society Instituted at Amsterdam in favour of Drowned Persons, 1767–1771*, G. Robinson, London, 1773.
Coleridge, S.T., *The Notebooks of Samuel Taylor Coleridge*, ed. K. Coburn, 5 vols, Routledge & Kegan Paul, London, 2002.
Coleridge, S.T., *Biographia Literaria*, ed. J. Engell and W. Jackson Bate, Princeton University Press, Princeton NJ, 1993.
Colman, G., *The Plays of George Colman The Elder*, ed. Kalman A. Burnim, 6 vols, Garland Publishing, New York and London, 1983.
Craciun, Adriana, *Writing Arctic Disaster: Authorship and Exploration*, Cambridge University Press, Cambridge, 2016.
Crook, N., and Derek Guiton, *Shelley's Venomed Melody*, Cambridge University Press, Cambridge, 1986.

Crouch, L.E., 'Davy's *A Discourse, Introductory to A Course of Lectures on Chemistry*: A Possible Scientific Source of *Frankenstein*', *Keats–Shelley Journal* 27, 1978, pp. 35–44.

Curry, J., *Observations on Apparent Death from Drowning, Hanging, Suffocation by Noxious Vapours, Fainting-Fits, Intoxication, Lightning, Exposure to Cold, &c., &c. and an account of the proper means to be employed for recovery*, 2nd edn, E. Cox and Son, London, 1815.

D'Arcy Wood, G., *Tambora: The Eruption that Changed the World*, Princeton University Press, Princeton NJ and Oxford, 2014.

Darwin, E., *The Temple of Nature; or, the Origin of Society: A Poem, with Philosophical Notes*, J. Johnson, London, 1803.

Darwin, E., *The Botanical Garden, 1791*, 2 vols, Scholar Press, Menston, 1973.

Davy, H., *The Collected Works of Humphry Davy*, ed. John Davy, 9 vols, Smith, Elder and Co., London, 1839.

Davy, H., *Elements of Agricultural Chemistry*, 2nd edn, London, Longman, Hurst, Rees, Orme, & Browne, 1814.

Davy, J., *Memoirs of the Life of Sir Humphry Davy*, 2 vols, Longman, Rees, Orme, Brown, and Green, London, 1836.

[D'Oyly, G.], 'An Enquiry into the Probability of Mr. Hunter's Theory of Life …', *Quarterly Review* 43, 1819, pp. 1–34.

Eisenberg, M.S., *Life in the Balance: Emergency Medicine and the Quest to Reverse Sudden Death*, Oxford University Press, Oxford and New York, 1997.

Franklin, B., 'A letter from Mr. Franklin to Mr. Peter Collinson, F.R.S. concerning the effects of lightning', *Philosophical Transactions* 47, 1752, pp. 289–91.

Franklin, B., 'A letter of Benjamin Franklin, Esq; to Mr. Peter Collinson, F.R.S. concerning an electrical kite', *Philosophical Transactions* 47, 1752, pp. 565–7.

Frayling, C., *Frankenstein: The First Two Hundred Years*, Real Art Press, London, 2018.

Fulford T., and Debbie Lee and Peter J. Kitson, *Literature, Science and Exploration in the Romantic Era: Bodies of Knowledge*, Cambridge University Press, Cambridge, 2004.

Godwin, W., *William Godwin's Diary*, Oxford, Bodleian Library, MS. Abinger, e. 17, 20.

Godwin, W., *Lives of the Necromancers*, Frederick J. Mason, London, 1834.

[Godwin, W.,], *Report of Dr. Benjamin Franklin, and Other Commissioners, Charged by the King of France, with the Examination of Animal Magnetism, as now Practised at Paris. Translated from the French. With an Historical Introduction*, J. Johnson, London, 1785.

Goodfield-Toulmin, June, 'Some Aspects of English Physiology: 1780–1840', *Journal of the History of Biology* 2, 1969, pp. 283–320.

Gowing, L., 'Greene, Anne (*c.* 1628–1659)', *ODNB*.
Graham, J., *The General State of Medical and Chirurgical Practice, exhibited, shewing them to be Inadequate, Ineffectual, Absurd and Ridiculous*, 4th edn, R. Cruttwell, Bath, 1778.
Grinfield, E., *Cursory Observations on the 'Lectures on Physiology, Zoology, and the Natural History of Man, delivered at the Royal College of Surgeons by William Lawrence F. R. S. Professor of Anatomy and Surgery to that College, &c. &c. &c.' In a series of letters addressed to that Gentleman; with a Concluding Letter to his Pupils*, 2nd edn, T. Cadell and W. Davies, London, 1819.
Harkup, K., *Making the Monster: The Science Behind Mary Shelley's Frankenstein*, Bloomsbury, London, 2018.
Hay, D., *The Making of Mary Shelley's Frankenstein*, Bodleian Library, Oxford, 2019.
Higgins, D., *British Romanticism, Climate Change, and the Anthropocene: Writing Tambora*, Palgrave Macmillan, Cham, Switzerland, 2017.
Hindle. M., 'Vital matters: Mary Shelley's *Frankenstein* and Romantic science', *Critical Survey* 2, 1990, pp. 29–35.
Hogg, T.J., *The Life of Percy Bysshe Shelley*, 2 vols, Edward Moxon, London, 1858.
Home, E., 'Hints on the Subject of Animal Secretions', *Philosophical Transactions* 99, 1809, pp. 385–91.
Hunter, J., *A Treatise on the Blood, Inflammation, and Gun-shot Wounds...*, George Nicol, London, 1794.
Hunter, J., 'Proposals for the Recovery of People Apparently Drowned', *Philosophical Transactions* 66, 1776, pp. 412–25.
Jeaffreson, J.C., *The Real Shelley*, 2 vols, Hurst and Blackett, London, 1885.
Keats, J., *John Keats: The Major Works*, ed. Elizabeth Cook, Oxford University Press, Oxford, 2008
Kite, C., *Essay in the Recovery of the Apparently Dead*, C. Dilly, London, 1788.
Knapp, A., 'George Forster', *The New Newgate Calendar...*, 5 vols, J. Robins and Co., London, 1826, vol. 4, pp. 182–9.
Knellwolf, C., and Jane Goodall (eds), *Frankenstein's Science: Experimentation and Discovery in Romantic Culture, 1780–1830*, Ashgate, Aldershot, 2008.
Lawrence, W., *Lectures on Physiology, Zoology, and the Natural History of Man*, W. Callow, London, 1819.
Lawrence, W., *An Introduction to Comparative Anatomy and Physiology...*, J. Callow, London, 1816.
Levere, T. *Poetry Realized in Nature*, Cambridge University Press, Cambridge, 1981.
Linebaugh, P., 'The Tyburn Riots Against the Surgeons', in Douglas Hay, Peter

Linebaugh, John G. Rule, E.P. Thompson and Cal Winslow (eds), *Albion's Fatal Tree: Crime and Society in Eighteenth-Century England*, Penguin, Harmondsworth, 1973, pp. 65–117.

Macilwain, G., *Memoirs of John Abernethy*, 3rd edn, Hatchard, London, 1856.

Marshall, T., *Murdering to Dissect: Grave-Robbing, 'Frankenstein', and the Anatomy of Literature*, Manchester University Press, Manchester, 1995.

Medwin, T., *The Life of Percy Bysshe Shelley*, 2 vols, Thomas Cautley Newby, London, 1847.

Mellor, A.K., 'Frankenstein: A Feminist Critique of Science', in George Levine (ed.), *One Culture: Essays in Science and Literature*, University of Wisconsin Press, Madison WI, 1987, pp. 287–312.

Mercer, A., *The Collaborative Literary Relationship of Percy Bysshe Shelley and Mary Wollstonecraft Shelley*, Routledge, London, 2019.

Morus, I.R., *Shocking Bodies:* Life, Death & Electricity in Victorian England, History Press, Cheltenham, 2011.

Morus, I.R., *Frankenstein's Children*, Princeton University Press, Princeton NJ, 1998.

Mulvey Roberts, M., 'The Corpse in the Corpus: *Frankenstein*, Rewriting Wollstonecraft and the Abject', in Michael Eberle-Sinatra (ed.), *Mary Shelley's Fictions*, Macmillan, London, 2000, pp. 197–210.

Peake, Richard Brinsley, *Presumption; or, the Fate of Frankenstein*, ed. Stephen C. Behrendt, 2001, https://romantic-circles.org/editions/peake/index.html.

Phillipe A., *The Hour of Our Death*, trans. Helen Weaver, Allen Lane, London, 1981.

Polidori, J., *The Diary of John William Polidori*, ed. William Michael Rossetti, Elkin Mathews, London, 1911.

Priestley, J., *Experiments and Observations on Different Kinds of Air*, 6 vols, J. Johnson, London, 1774–86.

Rennell, T., *Remarks on Scepticism, Especially as it is Connected with the Subjects of Organization and Life. Being an Answer to the Views of M. Bichat, Sir T.C. Morgan and Mr. Lawrence upon these Points*, F.C. and J. Rivington, London, 1819.

Resuscitation Council UK, www.resus.org.uk/resuscitation-guidelines/introduction.

Richardson, Ruth, *Death, Dissection and the Destitute*, 2nd edn, University of Chicago Press, Chicago IL, 2001.

Robinson, C.E., *The Frankenstein Notebooks: A Facsimile Edition of Mary Shelley's Manuscript Novel, 1816–17*, 2 vols, Garland Publishing, New York, 1996.

Ruston, S., *Creating Romanticism: Case Studies in the Literature, Science and Medicine of the 1790s*, Palgrave Macmillan, Basingstoke, 2013.

Ruston, S., *Shelley and Vitality*, Palgrave Macmillan, Basingstoke, 2005.

Shelley, M., *Frankenstein: The 1818 Edition with Related Texts*, ed. David Wootton, Hackett Publishing, Indianapolis IN and Cambridge, 2020.

Shelley M., *Frankenstein: 1818 Text*, ed. Nick Groom, Oxford University Press, Oxford, 2018.

Shelley, M., *The New Annotated Frankenstein*, ed. Lesley S. Klinger, Liveright, New York, 2017.

Shelley, M., *Frankenstein*, ed. M.K. Joseph, Oxford University Press, Oxford, 1968, repr. 2008.

Shelley M., *The Journals of Mary Shelley: 1814–1844*, ed. Paula R. Feldman and Diana Scott-Kilvert, Johns Hopkins Press, Baltimore MD and London, 1995.

Shelley, M., *The Letters of Mary Wollstonecraft Shelley*, ed. Betty T. Bennett, 2 vols, Johns Hopkins University Press, Baltimore MA, 1980.

Shelley M., and P.B. Shelley, *History of a Six Weeks' Tour*, Woodstock Books, Oxford, 1989.

Shelley, P.B., *The Poems of Shelley*, ed. Geoffrey Matthews, Kelvin Everest, Jack Donovan, Cian Duffy and Michael Rossington, 4 vols, Longman, London and New York, 1989–2013.

Shelley, P.B., *The Witch of Atlas Notebook: A Facsimile of Bodleian MS. Shelley adds. e. 6*, ed. Carlene A. Adamson, Garland Publishing, New York, 1997.

Shelley, P.B., *Miscellaneous Poetry, Prose and Translations from Bodleian MS. Shelley adds. c.4, Folios 63, 65, 71, and 72*, ed. E.B. Murray, Garland Publishing, New York, 1992.

Shelley, P.B., *Shelley's Poetry and Prose*, ed. Donald H. Reiman, W.W. Norton, New York, 1977.

Shelley, P.B., *Shelley's Prose, or the Trumpet of a Prophecy*, ed. David Lee Clark, University of New Mexico Press, Albuquerque NM, 1966.

Shelley, P.B., *The Letters of Percy Bysshe Shelley*, ed. F.L. Jones, 2 vols, Clarendon Press, Oxford, 1964.

Smellie, W., *The Philosophy of Natural History*, C. Elliot, T. Kay, T. Cadell, and G.G.J. & J. Robinsons, Edinburgh and London 1790.

Smith, A., 'Scientific Contexts', *The Cambridge Companion to Frankenstein*, ed. Andrew Smith, Cambridge University Press, Cambridge, 2016.

Thornton, J.L., 'John Abernethy 1764–1831', *St Bartholomew's Hospital Journal* 72, July 1964, pp. 287–93.

Thornton, J.L., *John Abernethy: A Biography*, Printed by the Author, London, 1953.

Thornton, R., *The Philosophy of Medicine: being Medical Extracts on the Nature and Preservation of Health, and on the Nature and Removal of Disease*, 2 vols, 5th edn, Sherwood, Neely & Jones, London, 1813.

Tims, M., *Mary Wollstonecraft: A Social Pioneer*, Millington, London, 1976.

Turney, J., *Frankenstein's Footsteps: Science, Genetics and Popular Culture*, Yale University Press, New Haven CT, 1998.

Ure, A., 'An Account of Some Experiments made on the Body of a Criminal immediately after Execution, with Physiological and Practical Observations', *Quarterly Journal of Science* 6, 1819, pp. 283–94.

Walker, A., *Analysis of a Course of Lectures on Natural and Experimental Philosophy*, 14th edn, J. Barfield, London, 1807.

Walker, A., *A System of Familiar Philosophy: In Twelve Lectures; Being the Course of Lectures Usually Read by Mr. A. Walker*, 2nd edn, 2 vols, G. Kearsley, London, 1802.

Wesley, J., *The Desideratum: Or, Electricity made Plain and Useful by a Lover of Mankind and of Common Sense*, W. Flexney, London, 1760.

[Whewell, W.], '*On the Connexion of the Physical Sciences*. By Mrs. Somerville', *Quarterly Review* 51, 1834, pp. 54–68.

Williams, C., '"Inhumanly Brought Bach to Life and Misery": Mary Wollstonecraft, *Frankenstein*, and the Royal Humane Society', *Women's Writing*, vol. 8, no. 2, 2001, pp. 213–34.

Wollstonecraft M., *The Collected Letters of Mary Wollstonecraft*, ed. Janet Todd, Allan Lane, London, 2003.

Wollstonecraft, M., *The Works of Mary Wollstonecraft*, ed. Janet Todd and Marilyn Butler, 7 vols, William Pickering, London, 1989.

Wordsworth, W., *The Prose Works of William Wordsworth*, ed. W.J.B. Owen and J.W. Smyser, 3 vols, Clarendon Press, Oxford, 1974.

Wright, A., *Mary Shelley*, University of Wales Press, Cardiff, 2018.

Wright, A., and Dale Townshend, *Romantic Gothic: An Edinburgh Companion*, Edinburgh University Press, Edinburgh, 2015.

PICTURE CREDITS

COLOUR PLATES

PLATE 1 © Bodleian Library, University of Oxford, Arch. AA e.167 (v.1), title page
PLATE 2 © Bodleian Library, University of Oxford, Shelley relics (d)
PLATE 3 © Bodleian Library, University of Oxford, Shelley relics 15
PLATE 4 © National Portrait Gallery, London
PLATE 5 © Bodleian Library, University of Oxford, MS. Abinger, d. 27, fol. 55r
PLATE 6 © Bodleian Library, University of Oxford, MS. Abinger c. 56, fol. 12r
PLATE 7 © Bodleian Library, University of Oxford, MS. Abinger c. 56, fol. 12v
PLATE 8 Yale University Art Gallery
PLATE 9 Birmingham Museums Trust, 1905P2
PLATE 10 Wikimedia Commons
PLATE 11 Collection of Derby Museums
PLATE 12 © Ian Dagnall/Alamy Stock Photo
PLATE 13 © Bodleian Library, University of Oxford, MS. Shelley adds. e. 6, pp. 170–71
PLATE 14 Wellcome Collection
PLATE 15 © History of Science Museum, University of Oxford, Inv. 60861
PLATE 16 Wellcome Collection
PLATE 17 Wellcome Collection
PLATE 18 © History of Science Museum, University of Oxford, Inv. 70367
PLATE 19 Philadelphia Museum of Art: Gift of Mr and Mrs Wharton Sinkler, 1958, 1958-132-1

PLATE 20   © Bodleian Library, University of Oxford, Curzon b.3(174)
PLATE 21   © Bodleian Library, University of Oxford, Curzon b.30(47)
PLATE 22   Wellcome Collection
PLATE 23   © Bodleian Library, University of Oxford, Per. 1918 d. 127, vol XI, fig. 941
PLATE 24   © Bodleian Library, University of Oxford, MS. Abinger e. 21, fol. 13v
PLATE 25   © Bodleian Library, University of Oxford, MS. Shelley, c. 1, fols 212v–213r
PLATE 26   © Bodleian Library, University of Oxford, MS. Abinger c. 56, fol. 21r
PLATE 27   Wellcome Collection
PLATE 28   Wellcome Collection
PLATE 29   © Bodleian Library, University of Oxford, Curzon b.4(83)
PLATE 30   Wellcome Collection
PLATE 31   Wellcome Collection
PLATE 32   Wellcome Collection

FIGURES

FIG. 1   National Gallery of Art, Washington, Gift of Robert S. Pirie, 1981.77.3
FIG. 2   Wellcome Collection
FIG. 3   Yale Center for British Art, Paul Mellon Collection, B1977.14.14464
FIG. 4   Wellcome Collection
FIG. 5   Wellcome Collection
FIG. 6   Wellcome Collection
FIG. 7   Wellcome Collection
FIG. 8   Wellcome Collection
FIG. 9   Wellcome Collection
FIG. 10 Wellcome Collection
FIG. 11 Houghton Library, Harvard University, f Typ 815.67.3922
FIG. 12 Oxford Museum of Natural History / Scott Billings
FIG. 13 Wellcome Collection
FIG. 14 Wellcome Collection
FIG. 15 Wellcome Collection
FIG. 16 Wellcome Collection

INDEX

References to illustrations are in *italic* type

Abernethy, John *86*
    debate with William Lawrence
        81–2, 85–8, 100–103, 105
    on the vital principle 7, 90–95
    Surgical Observation on the
        Constitutional Origin and
        Treatment of Local Diseases 87
Agrippa, Heinrich Cornelius 12–13
air pump 35, 41–3, 66
alchemy 12–13, 44, 66, 108
Aldini, Giovanni 70–72, 74–6
anatomical dissection 38, 54, 70–73
anatomy, comparative 46, 93, 96–8
Anatomy Act 120
Animal Chemistry Society 110, 112
animal electricity *see* electricity
animal magnetism 33–4, 77–8
animal rights 22, 58
    see also vivisection
artificial breathing 52–3
artificial intelligence 121
asphyxiation 53
    by hanging 54, 56, 73
    strangulation 51, 53, 56, 59
    suffocation 27, 50–51, 74
atmospheric electricity *see* electricity

automata 30, 32–5

Barbauld, Anna Letitia
    'A Mouse's Petition' 20–22, 42
    'Life' 22
Bichat, Marie François Xavier 37,
    100–101, 109
    Physiological Researches on Life and
        Death 95–6
Blizzard, Sir William 85
blood transfusion 31
Blumenbach, Johann Friedrich 24,
    83, 98
Brown, John 24
    Elements of Medicine 114–15
Buffon, Georges-Louis Leclerc,
    Comte de, *Histoire Naturelle* 12, 83
burial 117
    premature 7, 55
Burke, William, and William Hare
    119–20; *Plate 30*
Byron, Allegra 2
Byron, George Gordon, Lord 1, 26
    Don Juan 26–7
    Manfred 27; *Plate 8*
    'Prometheus' 28

carbon dioxide 47
Chain of Being 83–4
chemical instruments 15, 67
chemistry 13–15, 43–4, 64–6, 80, 112–13
Clairmont, Claire 1–2, 26
clean air 41–2, 122
    see also fresh air
Coleridge, Samuel Taylor 24, 83
    'The Eolian Harp' 22
    Hints towards the Formation of a more Comprehensive Theory of Life 25
    'Human Life, On the Denial of Immortality, a Fragment' 25–6
    Kubla Khan 34
    The Rime of the Ancient Mariner 23–4
Colman, George, *The Genius of Nonsense* 79
corpse 37, 43, 47, 108, 117–20
    galvanized 27, 62, 71–5
    rights of the corpse 117
Curry, James, *Observations on Apparent Death* 51–3, *53*, 56, 58

Darwin, Erasmus 33
    Temple of Nature 82
Davy, Sir Humphry 13–14, 55, 65–7, 80; Plate 4
    Discourse, Introductory to a Course of Chemistry 13
    Elements of Agricultural Chemistry 13, 46, 64, 113; Plate *13*
    Elements of Chemical Philosophy 13, 112
dead bodies *see* corpses
death
    incomplete and absolute 7, 51
    and life 47, 100, 108–13
dreams 8, 34, 36, 38, 51
    nightmares 32–3, 107, 115

drowning 7–9, 48–51, 122–3

electricity 61–70, 92
    animal electricity 69, 93
    atmospheric electricity 61, 67–9, 77
    electric battery 60, 64–5, 67
    electric machine 35, 43, 61, 66–7
    electric shock 62, 77
    electrical resuscitation 75–6
    electrical therapy 76–7
electrochemistry 65–7

fainting 10, 32, 40, 116
fire air 44–5
Franklin, Benjamin 63, 75; Plate *19*
fresh air 41–2
    see also clean air

Galvani, Luigi 62, 69
galvanic trough 35, 64
galvanism 6, 27, 62, 72, 80
Godwin, William 8, 10, 12–13, 90, 124; Plate *24*
    Lives of the Necromancers 12
    Report of Benjamin Franklin 77
    St Leon 12
Graham, James 78–80; Plate *22*
    see also Colman, George
grave-robbing 7, 30, 38, 56, 105; Plate *31*
Greene, Anne 54
Grotta del Cane, Naples 57–8, *57*
guillotine 109

heated bath *49*
Hoffman, E. T. 33
Hogg, Thomas Jefferson 4, 64, 67
Hunter, John 51, 75, *85*, 102, 118
    A Treatise on the Blood, Inflammation, and Gunshot Wounds 90–91

imagination 24–5, 35
Imlay Godwin, Frances 2
Ingenhousz, Jan 46

Keats, John 37–9
   'Ode to a Nightingale' 38
   'This Living Hand' 39
Kite, Charles, *Essay in the Recovery of the Apparently Dead* 75–6

Lack, Henrietta 121
laudanum 2, 114–15
Lavoisier, Antoine 43–5, *43*
Lawrence, William 5, 82, 88–90, *89*, 96–102
   debate with Abernethy *see under* Abernethy
   Lectures on Physiology, Zoology, and the Natural History of Man 104–5
light 45–7
lightning *see* electricity
Luddite movement 33

machines 33, 121
   see also automata, electric machine
magnetism *see* animal magnetism
materialism, French 102–3
Mesmer, Franz Friedrich Anton 33
Mesmerism 33–4
microscope 83–4
   solar microscope 64, 84; *Plate 15*
mimosa 84; *Plate 23*
Modern Sceptics 102–3
Morgan, Sir Thomas Charles 37–8

natural philosophy 12–15, 66
nervous illness 115–16

oneirodynia 32
   see also dreams, sleep, somnambulism
opium 24, 34, 114–15
   see also laudanum
Ovid, *Metamorphoses* 5
oxygen 21, 26, 40, 43–7, 52–3, 57
   see also fire air

'Philadelphia Experiment' 63
photosynthesis 45–7
Polidori, John 1, 6–7, 30–33, 39
   'The Vampyre' 31–2
premature burial *see* burial
Priestly, Joseph 20–22, 43
   discovery of oxygen 44
   photosynthesis 45
Prometheus 28, 62–3
public dissection 54

Rennell, Thomas, *Remarks on Scepticism* 104
reproduction 83
resurrection 19, 29, 72–4, 117
resurrection men 118
resuscitation 9–10, 123; *Plate 16*
Royal Humane Society 7–10, 40, 48–51, 53–5; *Plate 17*
Royal Society for the Prevention of Cruelty to Animals 58

Scheele, Karl Wilhelm 44
Schussele, Christian, *Prometheus Bound 18*
science
   of the mind 24–5
   pushing boundaries 29, 42, 62
Shelley, Clara Everina 2
Shelley, Harriet 2
Shelley, Mary 1–3; *Plate 2*
   death of child 1, 3–5, 8, 39; *Plate 5*
   education 10–11, 46–7
   Frankenstein; or the Modern Prometheus
      composition 2, 6,11, 19, 34; *Plate 26*

Creature, the 5, 12, 15–16, 21, 41, 45, 47, 56–9, 68, 71–3, 93, 107
Frankenstein, Victor 12–13, 23, 28–9, 39–41, 48, 55–7, 62–3, 65–71 81–2, 90, 93–4, 98–100, 107–8, 113, 115–17
Frankenstein, William 10, 56, 68, 100
'Frankenstein syndrome' 121
Mary's 1831 Introduction 27, 45, 72, 83, 115
in National Theatre 24
Percy Shelley's 'Preface' 98
Presumption; or, the Fate of Frankenstein 15
Waldman, Professor 13–15; Plates 6–7
'On Ghosts' 37
reading 11–13
Shelley, Percy Bysshe 1–2, 4–5, 46, 64, 67, 78; Plate 3; Plate 25
A Vindication of Natural Diet 37
Adonais 37
Alastor; Or, the Spirit of Solitude 36
death 2, 50
and *Frankenstein* 14, 65–6
'Hymn to Intellectual Beauty' 36–7
'The Magnetic Lady to her Patient' 33
'Mont Blanc' 37
'Mutability' 36
'On Life' 35–6
'On the Devil and Devils' 84
Prometheus Unbound 28
Queen Mab 8–9, 105
'The Sensitive Plant' 84
Shelley, William 1–2, 10
sleep 32–3

see also oneirodynia, somnambulism
Smellie, William, *Philosophy of Natural History* 11
somnambulism 32
soul 28, 34, 58, 92, 105
spontaneous generation 6, 82–3
suicide 2, 8, 48–9, 53
sunlight *see* light

technology 121–2
tobacco enema 52; Plate 20
trance 32, 38

vampirism 31
vegetarianism 37
Villa Diodati, Geneva 6
vision 19, 35
vivisection 20–21, 56–8, 97
see also animal rights
Volta, Alessandro 64
Voltaic pile *see* electric battery
von Kempelen de Pázmánd, Johann Wolfgang Ritter, 'The Turk' 33; Plate 10

Walker, Adam 64, 76–7
A System of Familiar Philosophy 77
Ward, William, *The Mouse's Petition* 21
'Waterloo Teeth' 118
weather 26, 67, 122
whaling 49–5; Plate 14
Wollstonecraft, Mary 1, 8, 11, 48, 124
A Vindication of the Rights of Woman 10–11, 77
review of William Smellie 11
Wollstonecraft Shelley, Mary *see* Shelley, Mary
Wright, Joseph, *An Experiment on a Bird in the Air Pump* 41–2, 56; Plate 12